Latin American Mirages, Mira[...]uth American Air Arms

Santi

LATIN AMERICAN MIRAGES

Mirage III/5/F.1/2000 in Service with South American Air Arms

Santiago Rivas and Juan Carlos Cicalesi

HARPIA
PUBLISHING+

Copyright © 2010 Harpia Publishing, L.L.C. & Moran Publishing, L.L.C. Joint Venture
2803 Sackett Street, Houston, TX 77098-1125, U.S.A.
lam@harpia-publishing.com

All rights reserved.

No part of this publication may be copied, reproduced, stored electronically or transmitted
in any manner or in any form whatsoever without the written permission of the publisher.

Consulting and inspiration by Kerstin Berger
Artwork and drawings by Tom Cooper and Ugo Crisponi
Editorial by Thomas Newdick
Layout by Norbert Novak, www.media-n.at, Vienna

Printed at Grasl Druck & Neue Medien, Austria

ISBN 978-0-9825539-4-7

Harpia Publishing, L.L.C. is a member of

Contents

Introduction 7

Acknowledgements 8

Abbreviations 9

Chapter 1 – Argentina 13
Mirage IIIEA/DA 13
Dagger and Finger 33
Mirage 5A Mara 78
Mirage IIIB/C 86

Chapter 2 – Brazil 99
Mirage IIIEBR/DBR 99
Mirage 2000C/B 111

Chapter 3 – Chile 115
Mirage 50 115
Mirage 5 Elkan 129

Chapter 4 – Colombia 137
Mirage 5COA/COR/COD 137
Kfir C2/TC2 and Mirage 5COAM/CODM 139

Chapter 5 – Ecuador 151
Mirage F.1JA/JE 151
Kfir C2/TC2 160
Mirage 50EV/DV 169
Atlas Cheetah C 170

Chapter 6 – Peru 173
Mirage 5P/DP 173
Mirage 2000P/DP 183

Chapter 7 – Venezuela 193
Mirage IIIEV and Mirage 5V/DV 193
Mirage 50EV/DV 198

Appendix I – Individual aircraft histories 209
Argentina
Mirage IIIDA 209
Mirage IIIEA 209
Mirage IIIBE 210
Dagger/Finger 210

Primary differences between Finger variants 211
Mirage 5A Mara 211
Mirage IIIB/C 212

Brazil
Mirage IIIDBR 213
Mirage IIIEBR 213
Mirage 2000C/B 214

Chile
Mirage 50 215
Mirage 5 Elkan 215

Colombia
Mirage 5COD/COR/COA 216
Kfir C2/C7/C10/C12/TC2/TC12 217

Ecuador
Mirage F.1JE/BE 218
Kfir C2/TC2 218
Mirage 50EV/DV 219
Cheetah C/D2 219

Peru
Mirage 5P/5DP 220
Mirage 2000P/DP 221

Venezuela
Mirage IIIEV 221
Mirage 5V/DV 221
Mirage 50EV/DV 222

Appendix II –
Camouflage patterns and primary armament of Latin American Mirages ... 223

Appendix III – Primary weapons used by Latin American Mirages 237

Appendix IV – Latin American Mirage units 239

Appendix V – Latin American Mirage family 245

Appendix VI – Kits and decals 247

Index 249

Introduction

Although the French aviation industry enjoyed a strong relationship with Latin America in the early years of aviation, after World War II the United States and United Kingdom took over as the primary suppliers of aircraft to the region, and particularly of combat types. The jet era began in Latin America with the supply of aircraft from both the US and UK. Among the first Latin American jet fighters were the Lockheed F-80 Shooting Star and North American F-86 Sabre from the US and the Gloster Meteor, Hawker Hunter, and de Havilland Vampire and Venom from the UK. In contrast, Marcel Dassault had little success exporting his early jet fighters to Latin America. Indeed, the only examples of early Dassault jet fighters to serve in Latin America arrived in the 1970s, in the form of the Ouragan and Super Mystère, for El Salvador and Honduras respectively. Both types were purchased second-hand from Israel.

By the mid-1960s, therefore, the air forces of Latin America were equipped with first-generation jet fighters, typified by the F-80 and Meteor, and these were proving inadequate for the requirements of the time. None of these aircraft were capable of supersonic speeds in level flight, and almost all lacked radar or missile armament. After 1965 most Latin American air arms began to look for new equipment, but encountered problems with their previous major suppliers. The US was engaged in the Vietnam War and was unwilling to sell new fighters to Latin America. Many Latin American countries attempted to acquire the Northrop F-5 Freedom Fighter, on account of its performance, low cost and simple operation. The F-5 may have looked like an ideal solution for Latin American air arms, but purchase requests were declined by the US – for the time being at least. The only country to acquire a significant quantity of combat aircraft from the US during this period was Argentina, with a batch of 50 Douglas A-4B Skyhawk fighter-bombers. The UK, meanwhile, was not in a position to offer any capable, modern fighters, the only British option, the English Electric Lightning, being considered obsolescent.

While Latin American air arms sought new fighter equipment, one event grabbed the attention of the world and had a significant impact on the staff of air forces in Latin America. From 5–10 June 1967, the Six-Day War saw Israel locked in combat with its Arab neighbours. In the air war, the undoubted 'star' of the conflict was the Mirage IIICJ serving with the Heyl Ha'Avir (Israel Defence Force/Air Force, IDF/AF).

At the time, the Mirage III was little known in Latin America, but local air forces soon learned that France was willing to sell the fighter to any customer. The Mirage was the aircraft that Latin America was looking for: capable of speeds of around Mach 2.2, and equipped with radar and missile armament, the French jet represented the latest technology and was available at an attractive price.

The first to sign a contract for the Dassault jet was Peru, in 1967. The Peruvian order concerned the Mirage 5, which was a new attack version of the Mirage IIIE developed for Israel, without radar and with two extra pylons below the fuselage. Two years later Peru's lead was followed by Argentina and Brazil, both of which purchased the Mirage IIIE, while in 1971 Colombia ordered the Mirage 5. Venezuela continued the trend in 1973 when it acquired a mix of Mirage IIIE and Mirage 5 jets. The success of the Mirage in the region continued when Argentina ordered a batch of IAI Neshers in 1978. These aircraft, essentially Mirage 5s assembled in Israel, were named Dagger in Argentine service. A new version of the Dassault fighter was introduced to Latin America by Ecuador from 1978, with the powerful Mirage F.1.

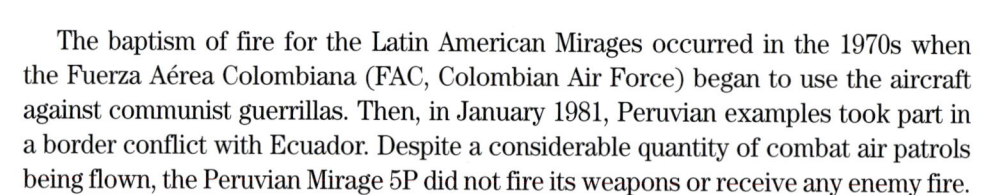

The baptism of fire for the Latin American Mirages occurred in the 1970s when the Fuerza Aérea Colombiana (FAC, Colombian Air Force) began to use the aircraft against communist guerrillas. Then, in January 1981, Peruvian examples took part in a border conflict with Ecuador. Despite a considerable quantity of combat air patrols being flown, the Peruvian Mirage 5P did not fire its weapons or receive any enemy fire.

Things would be very different in 1982, when Argentine fighters engaged British forces over the Malvinas/Falkland Islands. For the loss of 2 Mirage IIIEAs and 11 Daggers, the Argentine jets participated in the sinking of the British warship HMS *Ardent* and damaged HMS *Antrim*, HMS *Plymouth*, HMS *Brilliant*, RFA *Sir Bedivere*, RFA *Sir Lancelot* and RFA *Sir Tristram*.

Meanwhile, deliveries to the region continued, with Chile accepting the Mirage 50 in 1980 and Ecuador taking the new IAI Kfir C2 in 1981. During the Falklands War, Argentina received 10 Mirage 5s from Peru and, before the end of 1982, 23 Mirage IIIB/Cs had arrived in Argentina from Israel.

A major advance was made by Peru in 1986, with the purchase of 12 Mirage 2000s, the new model offering numerous improvements over the first-generation Mirages. These were followed by a delivery of Kfirs to Colombia.

The late 1980s and 1990s were times of modernisation, and projects of different levels were developed for the Mirage fleets in all countries in the region.

The final addition to the first-generation models was made by Chile, with its Mirage 5 Elkan, while the latest Latin American country to acquire Mirages was Brazil, recipient of 12 second-hand Mirage 2000C/Bs in 2005. As this book was being prepared, Chile and Venezuela were no longer operators of the family, and Argentina was looking for a replacement for its entire fleet, with the Mirage 2000 being the prime candidate. The story of the Mirage in Latin America is far from over, with a long life ahead for the Brazilian and Peruvian Mirage 2000 fleets, and for the Ecuadorian and Colombian Kfirs, while the story of the Latin American Cheetahs is just to start.

Santiago Rivas
Buenos Aires, June 2010

Acknowledgements

For their help in the preparation of this book, the author wishes to thank the following persons and institutions:
Gustavo Aguirre Faget, Christian Amado, Virgilio Aray, Oscar Arredondo, Atilio Baldini, Johnson Barros, Tsahi Ben-Ami, Heinz Berger, Roberto Bertazzo, Alessandro Bocca, César Bombonato, Mario Callejo, Aparecido Camazano Alamino, Eduardo Cardenas, Hernán Casciani, Guido Chávez Acosta, Horacio Clariá, Tom Cooper, Santiago Cortelezzi, Ugo Crisponi, César Cruz Tantaleán, Vicecomodoro Jose D'Andrea, Dassault, Brigadier Norberto Dimeglio, Major Fernando Estrella, FAA, FACh, FAC, FAE, FAV, Javier Franco 'Topper', Erwin Fuguet, Guillermo Galmarini, Suboficial Mayor Alfredo González, IAI, Patrick Laureau, Michel Liébert, Chris Lofting, Claudio Luchessi, Paulo Kasseb, Alberto Maggi, Comodoro Macaya, Vicecomodoro Maiztegui, Lewis Mejía, Joao Paulo Moralez, Thomas Newdick, Jaco du Plessis, Sergio de la Puente, Iván Peña Nesbit, Comodoro Carlos Eduardo Perona, Álvaro Romero, Stefano Rota, Mariano Salcedo, Amaru Tincopa, Katsuhiko Tokunaga/D.A.C.T., Erwin van Dijkman, Cees-Jan van der Ende, Elio Viroli, Rogier Westerhuis.

Abbreviations

1°GAvCa	1° Grupo de Aviação de Caça, 1st Fighter Aviation Group
1° GDA	1° Grupo de Defesa Aérea, 1st Air Defence Group
AAA	anti-aircraft artillery
ADC	air data computer
ADF	automatic direction finder
ALADA	1° Ala de Defesa Aérea, 1st Air Defence Wing
AMARC	Aerospace Maintenance and Regeneration Center at Davis-Monthan Air Force Base, United States
AMBV	Aviación Militar Bolivariana de Venezuela, Venezuelan Bolivarian Military Aviation
AMD	Avions Marcel Dassault
ANG	Air National Guard
APU	auxiliary power unit
ARA	Armada de la República Argentina, Argentine Navy
ARC	Armada de la República de Colombia, Colombian Navy
ARV	Armada de la República de Venezuela, Venezuelan Navy
ARMACUAR	Área de Material Río IV
Atar	Atelier Technique Aéronautique Rickenbach
AWACS	Airborne Warning And Control System
BAAN	Base Aérea de Anápolis, Brazil
BAAQ	Base Aeronaval Almirante Quijada, Río Grande, Argentina
BAAZ	Base Aeronaval Almirante Zar, Trelew, Argentina
BACE	Base Aeronaval Comandante Espora, Bahía Blanca, Argentina
BACO	Base Aérea Canoas, Brazil
BAC	British Aerospace Corporation
BAe	British Aerospace
BAM	Base Aérea Militar, Military Air Base
BAPI	Base Aeronaval Punta Indio, 120km (75 miles) southeast of Buenos Aires
BOAC	Base Oficial de Aviación Civil, Civil Aviation Official Base
BRP	Bomba retardada por paracaídas, parachute-retarded bomb
c/n	construction number
CACOM-1	Comando Aéreo de Combate 1, 1st Combat Air Command
CAP	combat air patrol
CC	Capitán de Corbeta
CdoFAS	Comando de la Fuerza Aérea Sur, Southern Air Force Command
CEASO	Centro de Ensayos de Armamentos y Sistemas Operativos, Operational Systems and Weapons Test Centre
CEPAI	Comissão de Estudos do Projeto da Aeronave de Interceptação, Interceptor Project Research Commission
CEV	Centre d'Essais en Vol, Air Test Centre
CIC	Centro de Información y Control, Control and Information Centre
CINDACTA-1	Centro Integrado de Defesa e Controle de Tráfego Aéreo, Air Traffic Control and Defence Integrated Centre

CLDS	Cockpit Laser Designator System
COAN	Comando de Aviación Naval, Argentine Naval Aviation
CODA	Centro de Operações de Defesa Aérea, Air Defence Operations Centre
COMAC	Comando Aéreo de Combate, Air Combat Command
COMDA	Comando de Defesa Aérea, Air Defence Command
DME	distance measuring equipment
DTC	data transfer cartridge
ECM	electronic countermeasures
Ejército del Aire	Spanish Air Force
ELINT	electronic intelligence
ELMA	Empresa Líneas Marítimas Argentinas, Argentine state-owned maritime company
ELN	Ejército de Liberación Nacional, National Liberation Army
EMD	Electronique Marcel Dassault
ENAER	Empresa Nacional de Aeronáutica
ESM	electronic support measures
EU	electronic unit
FAA	Fuerza Aérea Argentina, Argentine Air Force
FAB	Força Aérea Brasileira, Brazilian Air Force
FAC	Fuerza Aérea Colombiana, Colombian Air Force
FACh	Fuerza Aérea de Chile, Chilean Air Force
FAE	fuel-air explosive, a type of incendiary bomb
FAE	Fuerza Aérea Ecuatoriana, Ecuadorian Air Force
FAS	Fuerza Aérea Sur, Southern Air Force
FAP	Fuerza Aérea del Perú, Peruvian Air Force
FARC	Fuerzas Armadas Revolucionarias de Colombia
FAV	Fuerza Aérea Venezolana, Venezuelan Air Force
FFAR	Folding-Fin Aerial Rocket
FIDAE	Feria Internacional del Aire y el Espacio, Space and Air International Fair
FMA	Fábrica Militar de Aviones, the Military Aircraft Factory of the Argentine Air Force
FMS	Foreign Military Sales
GAv	Grupo de Aviação, Aviation Group
G1VAE	Grupo 1 de Vigilancia Aérea Escuela
GADA	Grupo de Artillería de Defensa Aérea, Air Defence Artillery Group
HMD	helmet-mounted display
HMS	Her Majesty's Ship
HOTAS	hands on throttle and stick
HUD	head-up display
IAI	Israel Aircraft Industries (now Israel Aerospace Industries)
IDF/AF	Israel Defence Force/Air Force (Heyl Ha'Avir)
IFF	identification friend or foe
ILS	instrumental landing system
INS	inertial navigation system
IR	infra-red

LCD	liquid-crystal display
LSL	Landing Ship Logistic
MER	multiple ejector rack
MirSIP	Mirage System Improvement Programme
MLU	mid-life upgrade
MPC	Mission Planning Centre
NATO	North Atlantic Treaty Organization
NuCOMDABRA	Núcleo do Comando de Defesa Aérea Brasileira, Brazilian Air Defence Command Core
OF	Orden Fragmentaria, literally Fragmentary Order, the order to fulfil a mission sent by Argentina's Comando de la Fuerza Aérea Sur to different flying units
OPO	Oficial de Permanência Operacional, Operational On Guard Officer
PAMA SP	Parque de Material Aeronáutico de São Paulo, São Paulo Aeronautical Material Facility
RAF	Royal Air Force
REI	Programa Reemplazo Equipamiento Inglés, British Equipment Replacement
RN	Royal Navy
ROA	Red de Observadores del Aire, Air Observers Net
RWR	radar warning receiver
SAM	surface-to-air missile
SAR	search and rescue
SEMAN	Servicio de Mantenimiento, Maintenance Service
SIGINT	signals intelligence
SINT	Sistema Integrado de Navegación y Tiro, Gunnery and Navigation Integrated System
SNCASE	Société Nationale de Constructions Aéronautiques du Sud-Est
SNECMA	Société Nationale d'Etudes et de Constructions de Moteurs d'Aviation
SISDACTA	Sistema de Defesa Aérea e Controle de Tráfego Aéreo, Air Traffic Control and Air Defence System
TACAN	Tactical Air Navigation
TER	triple ejector rack
TEZ	Total Exclusion Zone
TOAS	Teatro de Operaciones Atlántico Sur, South Atlantic Operations Theatre
UFCP	up-front control panel
UHF	ultra high frequency
USAF	United States Air Force
VHF	very high frequency
VLF	very low frequency
VTOL	vertical take-off and landing
WDNRS	Weapons Delivery Navigation and Reconnaissance System
YPF	Yacimientos Petrolíferos Fiscales, Argentine state-owned petroleum company, now owned by Repsol

Chapter 1

ARGENTINA

Mirage IIIEA/DA

In the mid-1960s, the Fuerza Aérea Argentina (FAA, Argentine Air Force) maintained around 20 Gloster Meteor F.Mk 4s for the air defence of Buenos Aires, the country's capital. These were operated by VII Brigada Aérea (VII Air Brigade) at Morón, in the suburbs of the city. The FAA was also equipped with around 25 North American F-86F Sabres operated by IV Brigada Aérea at Mendoza.

The Meteor and the Sabre were limited in terms of weapons, and both lacked radar, so interceptions had to be guided by the ground-based radar of Grupo I de Vigilancia Aérea Escuela (G.I.V.A.E.). Faced with the requirement to replace the ageing Meteors, the FAA began to study new equipment, including the North American F-100 Super Sabre and the McDonnell Douglas F-4 Phantom II. The F-100 was rejected on account of its relative age, while the US refused to sell Argentina the F-4.

At the same time, the English Electric Lightning, the Mirage III, the Saab 35 Draken and the Lockheed F-104 Starfighter were analysed. The Northrop F-5 was initially included among the options under study. Despite its cheaper price tag, the F-5 was soon turned down as a result of the clear superiority of the competing designs.

By 1968, Peru was receiving its first batch of Mirages, while Brazil, Venezuela and Colombia were all examining purchases of the fighter. These factors, together with the decisive participation of the Israeli Mirage III during the Six-Day War and a powerful campaign of lobbying by Avions Marcel Dassault led to the FAA's decision to purchase the aircraft.

The original plan called for around 100 Mirages to be built locally. The quantity was later reduced to 50, but with limited available budgets, the FAA abandoned the licence-production proposal in 1967 and instead signed a pre-contract for a batch of Mirage IIIE and Mirage IIID aircraft. These were to be completed without the Doppler radar, on the request of the FAA.

The final contract signing was also delayed by budgetary reasons. Finally, on 14 July 1970 contract MIII/70 was inked and was approved by the government under Decree No. 1710 of 14 October 1970. The contract covered the purchase of 20 Mirage IIIEs at a total cost of 9,992,700 Francs each and two Mirage IIIDs for 9,406,810 Francs each. All aircraft were to be newly built and delivered during 1972. Subsequently the quantity of Mirage IIIEs was reduced to 10. The single-seaters received the serial numbers I-003 to I-012, while the Mirage IIIDs were I-001 and I-002. The final contract was valued at 123,151,580 Francs, or approximately 28 million US Dollars, including training and

Latin American Mirages

Argentine Mirage IIIEA I-007 before delivery. (Dassault)

spares. The Argentine government also provided 21 million US Dollars to prepare an air base for the operation of the new jets.

In 1971 an FAA delegation travelled to Israel to obtain information on the use of the Mirage IIIC with the Heyl Ha'Avir. Meanwhile, two groups of personnel were charged with accepting the jets in France, while a cadre of pilots trained on the Mirage IIIB with Escadron de Chasse et de Transformation (ECT) 2/2 'Cotê d'Or' at Base Aérienne 102 Dijon-Longvic. After this initial stage, the pilots undertook operational training at Base Aérienne 132 Colmar-Meyenheim with Escadron de Chasse 1/13 'Artois'. Operational training was completed on the Mirage IIIE, and included both simulated air-to-air and air-to-ground combat. Meanwhile, three officers completed the instructor course at Dijon, in order to prepare the first Mirage pilots to be trained in Argentina.

While training was under way, the FAA was discussing options for basing the operational Mirages. After analysing a number of locations, in 1971 the Aerodrome Dr Mariano Moreno, part of the deactivated Base Oficial de Aviación Civil (BOAC, Civil Aviation Official Base) was selected. The base was located at José C. Paz, some 30km (19 miles) from downtown Buenos Aires. On 7 December 1971 Resolution 534/71 approved works to adapt the aerodrome. Two months later, in accordance with Day Order 2/72, the newly created Escuadrón Mariano Moreno took over all the former BOAC properties.

Mirage IIIEA I-003 is seen prior to delivery at Dassault's facilities. (Dassault)

Argentina

Seen here during Air Force Day in 1974, two-seat Mirage IIIDA I-001 was lost in an accident on 30 March 1979. (Dassault)

The Mirage arrives

The first Mirage IIIEA (I-003) was delivered in France on 21 August 1972 and arrived in Argentina on 23 September on board Lockheed C-130E Hercules TC-62. This fighter joined the first Mirage IIIDA, which had arrived on 5 September on board C-130E TC-63. Before the end of the year aircraft I-004 and I-005 had also arrived in Argentina.

The jets were assembled at the base with the support of Dassault technicians and did not fly until 10 January 1973, when AMD pilot Gerald Resal took to the air in I-003. Seven days later, I-001 completed the first flight with Argentine crew. The first pilot course began on 19 March, led by Argentine instructors together with a French major, and the first solo flight by an Argentine pilot in the country took place on 18 October.

Day Order 90/73 established base Aérea Militar (BAM) Mariano Moreno on 21 September 1973. Created at the same time were Escuadrón I de Caza Interceptora, the Escuadrón Técnico and the Escuadrón Base. The remaining aircraft were received during the course of the year, and training began for day and night interception and aerial combat.

The Mirage was the first FAA aircraft capable of exceeding Mach 1 in level flight and initiated the age of vertical air combat and autonomous interception, thanks to its

Mirage IIIEA I-006 displays the camouflage scheme used until the 1990s. (Dassault)

Thompson-CSF Cyrano II radar. The jets could be armed with Matra R.530 infra-red and semi-active radar-guided air-to-air missiles. The R.530 was Argentina's first air-to-air missile, and the only such weapon in FAA service to offer radar guidance.

To improve the operational levels of the FAA pilots, courses began on a simulator belonging to the Fuerza Aérea Venezolana (FAV, Venezuelan Air Force) in March 1975. CECAPEM 90 aerial targets were purchased, to be installed on the Alkan PM3 pylon carried in the ventral position.

On 9 December 1975, VIII Brigada Aérea was created, and on 5 January the Grupo de Operaciones 8 was formed (later designated Grupo 8 de Caza), including Escuadrón I. The Grupo Técnico 8 was formed from the basis of the previous Escuadrón Técnico.

On 23 March 1976 the first accident took place, when I-009 crashed at Gualeguaychú, in Entre Ríos province, the aircraft being a write-off.

To augment its fleet of Mirages, contract MIII/77 was signed on 27 May 1977, covering an additional seven Mirage IIIEAs at a total cost of 224 million Francs. There were plans to add yet more Mirages – to provide a total of around 100 – but these were frustrated by budgetary restraints. Instead, an offer was accepted covering the purchase of 26 IAI Neshers from the Heyl Ha'Avir. These aircraft were renamed Dagger in Argentine service. Although the target of 100 new-build Mirage IIIEs was not reached, through recourse to variants and second-hand equipment, the 100-mark was almost attained. Ultimately, between 1972 and 1982 the FAA acquired a total of 92 Mirage/Dagger fighters of different models, becoming the largest operator of the deltas in the Americas.

The Mirage IIIEAs of the second batch formed Escuadrón II of Grupo 8 de Caza and were equipped with TACAN in an installation on top of the fuselage, the same equipment later being installed on the first batch of aircraft. The second batch was also able to launch the Matra R.550 Magic AAM, although the FAA had no missiles of this type when the second batch of Mirages arrived. In 1978 trials were made with the Rafael Shafrir air-to-air missile, which had been received with the Dagger, but these proved inferior to the Magic and were not used operationally on the Mirage IIIEA/DA.

A 1978 dispute with Chile over the three islands in the Beagle Channel, in the extreme south of the continent, almost led to a war between the two countries.

Inspecting the radar of Mirage IIIEA I-011. The jet is armed with bombs, despite its primary role as an interceptor. (Author's archive)

Mirage IIIEA I-004 carries target-towing equipment under the fuselage and Peruvian-supplied fuel tanks. (Author's archive)

On 8 December Escuadrón I was deployed to BAM Comodoro Rivadavia. Designated as Escuadrón Mirage, the unit was equipped with five Mirage IIIEAs and one Mirage IIIDA. The dispute was finally resolved on 24 December and the aircraft had returned to their home base by the beginning of 1979.

To continue to increase the Mirage fleet and to replace I-001, lost in an accident on 30 March 1979, on 27 July of this year another contract for 35 million Francs was signed, covering one Mirage IIIBE modified to Mirage IIIDA standard (I-020).

In October the first aircraft from the second batch (I-013) arrived in Argentina and was delivered to Grupo 8 de Caza in December, following assembly at Área de Material Río IV, the main workshops of the FAA, in Córdoba province.

An additional two-seater was purchased when contract Biplaza 1/80 was signed on 29 February 1980. This saw another Mirage IIIBE brought to Mirage IIIDA standard at a cost of 38.16 million Francs. The aircraft received the serial number I-021. On 8 October the arrival of the second batch of Mirages was completed with the delivery of I-109. These followed an order for R.550 Magic missiles, which would arrive just before the 1982 Falklands War.

Mirage IIIEA in combat

On 2 April 1982 the Argentine Navy and thousands of troops landed on the Malvinas/Falkland Islands, an archipielago located around 480km (300 miles) from the coast of South America and in dispute between this country and the United Kingdom. With the intention of providing cover for the landing, three Mirage IIIEAs (I-011, I-016 and I-019) were deployed to Río Gallegos on 29 March, with a stopover at Comodoro Rivadavia. Captains Raúl Gambandé and Jorge Testa and 1st Lieutenant Carlos Perona commanded the detachment. I-014 and I-017 reinforced them the following day. Although they were on alert throughout the day on April 2, the lack of enemy aircraft over the islands rendered their intervention unnecessary.

With the decision to invade the islands, the FAA immediately began to study potential aerial combat with the Royal Navy's British Aerospace Sea Harrier FRS.Mk 1, as

Mirage IIIEA I-016 pictured at Río Gallegos during the Falklands War. The aircraft is armed with one R.530 and two R.550 Magic missiles. (Vicecomodoro Maiztegui)

Latin American Mirages

A Mirage takes off for a mission during the final days of the war, armed with a single R.530 missile and carrying Peruvian fuel tanks. (Vicecomodoro Maiztegui)

well as possible interception of the Royal Air Force's Avro Vulcan bombers. The main problem was the distance to the islands, which limited the Mirages to no more than five minutes of air-to-air combat at high altitude. Furthermore, the R.550 Magic and R.530 were inferior to the AIM-9L Sidewinder AAM provided to the British by the US.

The Mirage III was to be employed primarily for air defence, although air-to-ground armament was kept ready in bunkers, in case it was necessary to reinforce the dedicated attack aircraft. By the end of the war the FAA had studied the possibility of using the R.530 as an anti-ship missile, but ultimately it was decided to retain the Mirage III for the air superiority role, and especially to prevent a British air strike launched from Chile against Argentine bases.

To increase the time over the islands, the FAA also examined the possibility of operating the Mirage from BAM Malvinas, but the idea was discarded for a number of reasons. Most critically, the runway was too short and could not be extended in time. In addition, the taxiways and tarmac were too small to disperse the Mirages to protect them against air strikes, and it was not possible to provide the necessary quantities of fuel and armament to operate the jets throughout the war.

Mirage IIIEA I-018 prepares for take-off from Río Gallegos. (Vicecomodoro Maiztegui)

After the creation of Fuerza Aérea Sur, it was decided to establish two mobile squadrons with Mirage IIIEAs. On 5 April Mirages I-003, I-008, I-010, I-015 and I-018 were deployed to IX Brigada Aérea at Comodoro Rivadavia, to form the Escuadrón Mirage Comodoro Rivadavia. The unit was under the command of Major Páez, and was supported by Learjet LV-OAS. Meanwhile, Escuadrón Mirage Río Gallegos was formed at Río Gallegos, under the command of Major José Sánchez. Days later, Río Gallegos saw the arrival of I-005 and I-006 from VIII Brigada Aérea, while I-015 and I-018 were sent from Comodoro Rivadavia on 23 April.

The two-seat I-002 and single-seaters I-004, I-007, I-012 and I-013 remained at Moreno for the defence of Buenos Aires, but only three were in service. Mirages I-002 to I-012 could only use the R.530 missile, while I-013 to I-019 could also employ the R.550 Magic. The R.550 missiles purchased in 1980 arrived on 15 April, and work began to put these weapons into service and train the crews in their operation. Since the pilots knew a little about the missiles already, the training could be accelerated.

Throughout the war, the Mirages at Comodoro Rivadavia were held back to defend the mainland, while those at Río Gallegos were to be used to protect the skies over the islands. Since the R.530 was of limited effectiveness, the Mirages from the second batch saw more widespread use.

First blood

Just after dawn on 1 May, when the news of the Vulcan raid against BAM Malvinas was received, OF.1090 was issued to Río Gallegos, ordering Fiera flight to be scrambled. This formation comprised the Mirages of Major Sánchez and Captain Marcos Czerwinski, armed with two Magics each, and was tasked with providing air cover over the islands. The aircraft took off at 06.40 and at 07.30 were flying over Puerto Argentino/Stanley, but they failed to make contact with the ground controller. Although combat air patrols were being flown over the islands, neither British nor Argentine fighters detected each other's presence, and the Mirages returned to the mainland.

The Vulcan raid was followed at 08.30 by the bombing of BAM Malvinas by Sea Harriers, and later the Royal Navy warships HMS *Glamorgan*, *Alacrity* and *Arrow* approached the coast to bombard the Argentine positions. In response, a number of flights were sent to attack the ships and the Mirage IIIEA received the order to provide air cover.

At 08.59 two Mirages took off in order to fulfil OF.1093. I-019 was flown by Captain Gustavo García Cuerva and I-015 by 1st Lieutenant Carlos Perona. Using the callsign Tablón, each jet was armed with two Magics and one R.530 to cover the Topo flight of A-4B Skyhawks. Arriving in the vicinity of the Falklands, the Centro de Información y Control (CIC) on the islands ordered them to remain orbiting at 30,000ft (9,144m). As the result of an error, the CIC then sent the Skyhawk flight to intercept a CAP, when the A-4s were carrying only bombs. Comodoro Perona remembers that *'at that moment, Captain García Cuerva took the decision to take the CAP from the Skyhawks and asked the CIC to send us to the British aircraft'*. The CAP was formed by two 801 Naval Air Squadron Sea Harriers flown by Lieutenant Commander Ward and Lieutenant Watson. *'We accomplished our aim'*, says Perona, *'because when the CAP knew we were behind them, they changed their course'*. Perona reported that the British air-

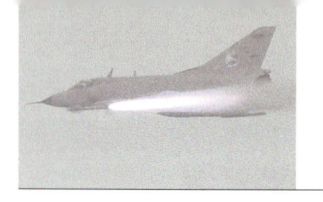

Mirage IIIEA I-014 displays the markings worn during the war. (Author's archive)

craft made a 180-degree turn to fly towards the Mirages, before the two flights crossed each other. *'We separated between 1,000 and 2,000ft (305 and 610m) to try and get a shot from the front with our R.530 missiles, but the radars did not have a good detection, and because of the low fuel we headed to Río Gallegos. We landed at 10.58 thinking we would not make it to base, as I only had 240 litres (53 Imp gal) of fuel and my boss had only 200 litres (44 Imp gal) when we landed.'*

At 10.23 another section of two Mirage IIIEAs took off with the callsign Foco (OF.1098), piloted by 1st Lieutenants Roberto Yebra (I-017) and Marcelo Puig (I-014). Their aim was to provide air cover to the Oso flight, but they failed to make contact with the Sea Harriers and returned to Río Gallegos. In the afternoon, the order to send two flights was given. The first was Buitre flight (fulfilling OF.1108) with Captain Gambandé (I-016) and 1st Lieutenant Yebra (I-014). After flying over the islands without finding any British aircraft, they landed at Río Gallegos at 17.50. Buitre flight was joined by Dardo flight (OF.1109), which took off at 15.45, comprising I-019 with Captain García Cuerva and I-015 with 1st Lieutenant Perona, armed with Magic missiles. Arriving in the combat zone, the CIC ordered the jets to orbit at 30,000ft (9,144m). García Cuerva asked about the condition of the runway at BAM Malvinas, since both pilots decided that, if they entered combat, they would jettison their external fuel tanks and then land at the airport to refuel. García Cuerva's message went unanswered.

Over Borbón/Pebble Island, Dardo flight entered combat with the 801 NAS Sea Harriers of Flight Lieutenant Paul Barton, RAF, and Lieutenant Steve Thomas. The Royal Navy fighters began a pincer movement. The Argentine pilots only saw Thomas's aircraft and Perona recalls that *'the radar operator decided to separate our flight to intercept the CAP, so we each turned 45 degrees to the opposite side of the heading we had. Then we ejected the 1,700-litre (374-Imp gal) tanks we were carrying, but a failure on my aircraft led to only one of them ejecting, and the right tank remained in place. Then the operator ordered another course change and told me that I had a target 30 miles (48km) ahead with an opposite heading and lower than me, so I started a shallow dive. Then they informed us that we were only 10 miles (16km) away. Because I didn't have the target on my radar I decided to make a visual search. I saw it by the contrast with the clouds I had below and when I was at about*

6 or 7 miles (10 or 11km) I reported that I had a Sea Harrier, on an opposing course [Thomas's aircraft]. *Before crossing I started a hard climb to gain height, but the tank I was still carrying reduced the climb performance and when I looked to one side I clearly saw the Sea Harrier climbing to my left, at about 500m (1,640ft), so we entered a scissors manoeuvre, reaching a separation of only 100 or 200 m (328 or 65ft)'.* At that moment, Barton, unseen by the Argentines, launched an AIM-9L missile against Perona's aircraft. Alerted by García Cuerva, Perona could not avoid the impact. He lost control of the Mirage immediately, while he watched the alarm lights activate. Perona informed García Cuerva that he would try to reach the coast, a distance of around 30 miles (48km), to avoid ditching in the water, since he was not wearing an anti-exposure suit. After flying without controls, and with the Mirage trying to roll because of the tank under the wing, Perona reached the coast and ejected. The pilot landed on Borbón/Pebble Island with only minor injuries. It was the Sea Harrier's first air-to-air victory of the war.

Thomas fired a missile against García Cuerva but missed, before the Argentine pilot evaded the British fighters in the clouds. With not enough fuel to return to the mainland, García Cuerva decided to attempt an emergency landing at BAM Malvinas. Major Silva from the Air Force Command Post on the islands instructed García Cuerva to eject, but he reported that his aircraft was in good shape and that he was confident of making a good landing. After some discussions, Silva informed the air defences that his aircraft was approaching.

At that moment the Falklands air defences were on alert after the Sea Harrier attack on the airport, and Silva told García Cuerva to use the air corridor provided for use by helicopters. García Cuerva approached low and at 180kt from over Sapper Hill to the northeast. Because of the risks of landing on a very short runway, he jettisoned his R.550 missiles. This action was misunderstood by an Argentine Army 35mm Oerlikon battery, which took the Mirage to be an attacking British aircraft, and opened fire in response. With his Mirage hit in the belly, García Cuerva shouted to Silva *'They are shooting at me!'* He made a right turn and headed south, before losing altitude and crashing off the coast south of Puerto Argentino/Stanley.

In the days that followed, the Mirage IIIEA remained on alert due to the possibility of an air strike against mainland bases by RAF Vulcans. Another threat was represented by the increased activity of Chilean combat aircraft and helicopters close to the border. The Mirages also made some patrol flights over Tierra del Fuego and Santa Cruz province, where it was feared that British special forces could be inserted to attack Argentine bases on the mainland. The Mirage IIIs were kept on five-minute alert and on several occasions were scrambled, specifically to intercept helicopters that appeared on Argentine radar, although the fighters never made contact with them.

New missions

After the actions on 1 May, the Mirage IIIEA never again entered combat with the Sea Harriers. The reason was not that the aircraft were not used again for CAP missions, rather that the Argentine interceptors always flew above 25,000ft (7,620m), while the attack aircraft flew much lower. The Sea Harriers never tried to hunt the Mirages, as they posed no direct threat to their operations against the Skyhawks and Daggers. If they flew lower, the Mirages would only have limited fuel with which to fight with the Sea Harriers, but they would have ensured that many more attack aircraft reached their targets without being intercepted.

The British landings on the islands took place on 21 May and the FAA launched numerous attack missions against the enemy ships. To cover these raids, two sections of Mirage IIIs were launched in the morning, beginning with Aguila flight (OF.1185) with Captain Jorge Huck and 1st Lieutenant Carlos Sellés, who took off at 09.56 armed with Magic missiles. The pair returned at 11.30 without having found any Sea Harriers. Meanwhile, Condor flight (OF.1186) took off at 10.01 with Major Sánchez and 1st Lieutenant Alberto Maggi at the controls. They returned at 11.40. Ciclón flight (OF.1200) was sent in the afternoon, with Major Marcos Czerwinski (I-017) and 1st Lieutenant Marcelo Puig (I-014) landing at 16.20.

Missions continued on 22 May, with Pitón flight (OF.1208), with Major Sánchez and Captain Czerwinski flying between 14.42 and 16.03. Then Cobra flight (OF.1209) took off, with Captains Huck and Guillermo Ballesteros completing their sortie between 15.45 and 17.22. Later that night, an unidentified air movement was detected close to Comodoro Rivadavia, and the Mirage of 1st Lieutenant Horacio Bosich was scrambled, armed with an infra-red R.530. FMA IA-58 Pucará A-558 flown by 1st Lieutenant Filipanics took off together with the Mirage. Bosich recalled this mission: *'The ground controller guided me to the target, until I made contact with the onboard radar. After tracking I prepared to fire, but I didn't because I wanted to be sure of the nature of the target. I tried to do a visual contact but it was impossible. I was guided four or five times to new targets, but without success. The operation was made at about 1,000ft (305m), over a terrain with hills, something that made it more complicated to know my real height.'*

Three missions were planned for 23 May, the first of which comprised Dardo flight (OF.1218) with Major Sánchez (I-005) and 1st Lieutenant Puig (I-026) armed with a single R.530 each. They flew their mission between 14.24 and 15.47. Flecha flight (OF.1219), with Captains Ballesteros (I-017) and Czerwinski (I-014) flew between 14.30 and 16.20. The day ended with a mission by Major Sánchez, callsign Ombú, at 16.33 to intercept a radar track detected 62 miles (100km) from Río Gallegos. The intruder was in level flight at 28,000ft (8,534m) and a speed of around 240kt, suggesting a helicopter. Sánchez failed to find the target and returned at 17.19.

Only one mission was launched on 24 May, by OF.1231, callsign Fuego. Armed with Magics, the section comprised Captains Huck and González, and the mission was planned to make possible the attack at San Carlos by Oro, Azul and Plata flights, by luring away the Sea Harrier CAPs. The Mirages took off for the joint diversion and reconnaissance mission with Pelo flight, which comprised Learjets T-21 and T-24 of the Fénix Squadron. At 10.55 the Mirages checked in with the CIC and were vectored to a Sea Harrier CAP that had been attracted by the Learjets, at a distance of 60 miles (97km). With the Argentine jets 30 miles (48km) away, the Sea Harriers made a turn and escaped, and the Mirages then returned to base at 12.00.

The following day one Mirage mission was launched, the aircraft commanded by Captain Ballesteros, with the callsign Patria (OF.1239). Ballesteros took off from Río Gallegos for an armed reconnaissance against a ship and a submarine spotted near the Argentine coast, while six Pucarás were prepared to attack the ships. After a low-level

I-014 photographed during a deployment to Patagonia. (Author's archive)

pass, Ballesteros reported that the ship was the YPF-owned (Yacimientos Petrolíferos Fiscales) tanker *Puerto Rosales*, refuelling the Argentine submarine ARA *San Luis*.

More missions were conducted in the following days, and on 26 May Sombra flight (OF.1242) took off at 13.32 with Captain Ballesteros and 1st Lieutenant Puig. Tasked with covering the Daggers of Poker flight, they returned at 15.20. On 27 May, Nene flight (OF.1252), comprising Captains Huck and González, took off at 16.35 to cover the A-4Bs of Truco flight, which bombed the refrigeration plant at Ajax Bay. The Mirages were each armed with two R.550s and one R.530. The jets landed at 18.20 after they had covered the Pucarás of Gaucho flight, which crossed from the mainland to BAM Malvinas, and a flight by C-130H TC-64. On 29 May, Captain Huck departed on a diversion mission with the callsign Gato (OF.2236), accompanied by two Learjets (callsigns Fuego and Tero), to cover a reconnaissance sortie by C-130H TC-64 with the callsign Loco.

Two days later, a plan was hatched under which Mirage IIIs were to fly over San Carlos, hoping that the Harriers would be scrambled from their forward operations base there. The intention was then to locate the airstrip and destroy it. OF.1271 called for the formation of Pitón flight, with I-014 and I-017 armed with R.530s and manned by Captains Ballesteros and Arnau. They took off at 06.10 but, with the radar on the islands put out of action by a combination of the RAF's Vulcan raid and AGM-45 Shrike missiles, the Mirages returned to base five minutes later. The same afternoon, Flecha flight (OF.1272) departed at 17.52 with the same intentions, and the same pilots, but after flying over the islands they failed to locate the runway at San Carlos.

The Mirage III did not fly again until 7 June, when Rayo flight took off to fulfil OF.1284, armed with R.530s and manned by Captain González and Major Sánchez. They were intended to cover Trueno flight, comprising A-4Bs on a mission to attack Mount Kent. The Mirages took off at 10.08, but returned to their base following the cancellation of the mission by Trueno flight. The offensive mission had been called off since the ground radar on the islands was out of action after an attack by Sea Harriers against a water tank, located only 300m (984ft) from the antenna.

On 8 June, during the attacks at Fitzroy, two Mirage missions were launched to provide air cover and diversion. At 16.00, Flecha flight took off according to OF.1300, the two aircraft armed with Magics and piloted by Captain Arnau and Major Carlos Luna. They were followed immediately by Lanza flight (OF.1301) with Captains González and Ballesteros. Flecha flight was informed that they had a Sea Harrier CAP on their tail at 12 miles (19km), and the Lanza section was sent to intercept the British jets. The Sea Harriers escaped the pursuers and Flecha flight finally landed at 17.40, followed by the Lanza section at 18.00.

Two days later, at 13.10, an FAA Boeing 707 conducting a surveillance flight spotted three aircraft, one large and two small, flying over the sea to the east of Buenos Aires, heading south. A Vulcan raid was suspected and at around 14.15 the bases in the south were alerted and four Mirage IIIs were scrambled. With OF.1306, Cóndor flight took off at 13.37 with Captains Arnau and González, armed with Magic missiles. They were followed by Daga flight (OF.1307) with Major Sánchez and Captain Ballesteros, who took off at 13.54. The fighters did not find the possible Vulcan, but they were alerted to the presence of two CAPs over Mount Kent and San Carlos. Cóndor flight landed at 15.18, followed at 15.37 by Daga flight. The 'CAPs' were in fact two RAF BAe Harrier GR.Mk 3s on a ferry flight from Ascension Island to the carriers, accompanied by a Victor tanker.

The Mirage's last combat sorties took place on 13 June, beginning with Triton flight, with Captains Arnau and Ballesteros. The aircraft took off at 15.13 from Río Gallegos, armed with two R.550s, to cover the Zeuz and Vulcano flights of Daggers, flying from Base Aeronaval Almirante Quijada in Río Grande. However, the Mirages were unable to intercept the Sea Harriers that tried to attack the Daggers and were forced to return.

The next mission was the last of the campaign for the Mirages and involved provision of escort for the final attack mission, conducted by the two BAC Canberras of Baco flight, which were to bomb the British forward positions. An escort was considered necessary as a result of the presence of CAPs on the previous Canberra bombing missions. Although the CAPs had not downed any Canberras, they were considered a serious threat and led to the cancellation of some missions.

In order to fulfil OF.1327, Plutón flight took off at 21.50, with the Mirages of Major Sánchez and Captain González, armed with Magics. After liaising with the ground controller, they were informed that no British air activity had been detected, so they stayed to the left of the Canberras until they dropped their bombs. The Mirage pilots watched the British field artillery fire on the Argentine positions, and the night fighting to the west of Puerto Argentino/Stanley, on Mounts Tumbledown and Longdon. The ground controller then informed the Mirages that missiles had been fired against them. The Mirage pilots claimed to see five missiles, but it is possible that bomb explosions confused them, as only one Sea Dart was fired by HMS *Cardiff*. Sánchez also reported that a missile passed very close to his Mirage. The Sea Dart hit Canberra B-108, while the radar operator informed of a CAP approaching at 80 miles (129km). The Mirages descended to escape the missiles and González ended his evasive manoeuvres over Goose Green at very low level, within the range of the British anti-aircraft artillery and missiles. Despite this close call, the Mirages returned safely to Río Gallegos shortly after midnight.

During the war, the Mirage IIIEA executed 45 combat air patrol and interception missions over the islands and 46 over the mainland and territorial waters, losing two aircraft and one pilot. After the loss of García Cuerva, most of the Mirages received yellow identification markings on the tail and wings. The two-seater I-002 also received these markings, although it was not used in combat.

After the battle

In November 1982 Mirage IIIDA I-020 was enlisted, following its arrival in the country in April. On 20 December Mirage IIIDA I-021, which had arrived in July, was also inducted.

On 1 January 1983 the Mirage IIIEA reached the milestone of 20,000 flying hours, and in June of the same year there were discussions concerning the purchase of more Mirage IIIs, although no orders were placed.

By 1983 Argentina had only 19 R.530 missiles and 20 R.530EM radar-homing seeker heads, 12 R.530IR infra-red seekers, 22 R.550 Magics and 18 Cyrano II radars. Twenty R.530s and 15 R.530 infra-red noses arrived together with numerous R.550 Magics from Libya during the Falklands War, these being transported onboard an FAA Boeing 707, together with many other spares for the Mirages. In order to return missile stocks to their original quantities, an additional batch of 28 R.550s and 9 R.530s were purchased direct from Matra in 1983.

By 1983 the number of Mirages in service had been reduced to 17, with an annual assignment of 3,000 hours. Since the early 1980s, the Fábrica Militar de Aviones (FMA, the FAA-owned aircraft industry) had been studying an indigenous replacement for the Mirage, under the designation SAIA 90, but the idea was abandoned shortly after Raúl Alfonsín became president in December 1983, precipitating a series of large-scale budget cuts for Argentina's armed forces.

After the war, the first batch of aircraft delivered received a series of modifications, comprising installation of Magic missiles (A-112), installation of VOR, ILS and Marker (A-116) and installation of DME (A-117). In 1985 Mirage I-014 was modified to serve as testbed for a Dassault proposal to add a HUD with ranging capability, and tailored for

Launching a Magic missile. (VI Brigada Aérea)

A Mirage after the war, with serial numbers removed. (Author's archive)

Latin American Mirages

A two-seater armed with two Magic missiles at Mendoza in the late 1980s.
(Author's archive)

the Magic as opposed to the R.530. The trigger was altered to separate firing of guns and missiles, and engine power reduction during firing was modified to suit these changes.

The FAA requested replacement of the Cyrano II radar, which proved very vulnerable to ECM, as did the radar-guided R.530 missile. At the same time the paint scheme was changed from the Southeast Asia pattern to air superiority grey. I-003 received an experimental scheme, but because of an error mixing the paint, the colour on the upper part of the aircraft was light blue. The jet retained this one-off scheme for some time until receiving the definitive colours. This process of repainting ended in 1991, when I-017 and I-018 received the new scheme.

Assigned hours for 1985 amounted to 27,000, 574 less than in 1984, while the number of aircraft in service remained the same. In March 1985 approval was given to OCH Industrial, in Cañada de Gómez, Santa Fe province, to acquire the tooling needed to manufacture canopies for the Mirage III and Dagger. Some time later, spares for the DEFA guns, internal fuel tanks and brakes also began to be built locally.

Mirages escort a Peruvian Air Force Fokker F28 carrying the Peruvian president on a visit to Argentina.
(FAA)

Argentina

I-003 displays an experimental paint scheme.
(Author's archive)

During 1986, air-to-air combat exercises took place among the various FAA units, with aircraft operating from the runway belonging to the Área de Material Río IV maintenance facilities. Intended to compare the relative performance of different types in light of lessons learned in the 1982 war, the exercises involved A-4B/C, Mirage IIIC, Mirage IIIEA and F-86F aircraft. In the same year, a modernisation proposal was issued to the FAA staff, as the Mirage was by now showing its age.

Activities in 1987 included a flight of Mirage IIIEAs escorting the aircraft carrying Pope John Paul II on 6 April, while on 19 August the entire Mirage IIIEA/DA fleet took part in the Air Force Day flypast.

VI Brigada Aérea

In 1987 it was decided to move the Mirage IIIEA and Mirage IIIDA to VI Brigada Aérea, at Tandil, where they would join the IAI Daggers and Fingers. The decision was dictated by the growth of the city around Moreno, which meant that Mirage operations were becoming more dangerous to the local population. By 7 March 1988 the aircraft had arrived at Tandil where they formed Escuadrón II of VI Brigada Aérea.

On 12 September 1990 a contract was signed with Spanish company Ceselsa to modernise 12 Mirage IIIEAs. Another contract was brokered with Intelsym to overhaul

A Mirage takes off during Exercise Aguila I at VI Brigada Aérea.
(Santiago Rivas)

27

Latin American Mirages

Escorting an Air France Concorde during the second visit of the airliner to Argentina in May 1978. The Concorde was carrying the French football team for the 1978 FIFA World Cup, held in Argentina in June that year.
(Author's archive)

and upgrade 12 Cyrano II radars, but the commission charged with evaluating the project did not approve the contracts.

As a result of significant budget cuts, only 925 hours were flown in 1991, with only 5 aircraft in service from a total of 15 Mirage IIIEA/DA jets listed in the inventory.

In 1992 the FAA began to study the possibility of replacing all its combat aircraft (A-4B/C, Mirage III and 5 and Finger), requesting permission from the US to buy McDonnell Douglas F/A-18 Hornet or General Dynamics F-16 Fighting Falcon fighters. Washington's response was a definitive 'no' in the case of the Hornet, while Argentina was informed that the F-16 was not to be made available in the short term. In the event, only the Skyhawks were replaced, with A-4Ms upgraded to A-4AR Fightinghawk standard.

As a stopgap measure, Argentina examined the possibility of purchasing 10 Mirage IIIEE and 4 Mirage IIIDE aircraft due to be retired by Spain, contracting the CASA-Ceselsa consortium to modernise these aircraft. Again, a lack of finances led to the cancellation of this project.

Intent on improving the capabilities of the fleet, an aerial refuelling probe was tested on a Mirage IIIEA in 1995, the aircraft being equipped with a Finger nose, but work on the project was abandoned some time later.

To mark the 25th anniversary of the Mirage, in 1997 serial number I-006 received a commemorative paint scheme.

The first multinational exercise in which the Mirage IIIEA fleet participated was Águila 1. On 14 August 1998, four Mirage IIIEAs and four Mirage 5s deployed to Villa Reynolds, home of V Brigada Aérea, where they joined the OA/A-4AR jets of that unit,

A Mirage III, with the nose of a Finger, was used to test a refuelling probe in the early 1990s. The modification was not implemented.
(VI Brigada Aérea)

Three Mirage IIIEA fighters escort Boeing 707 VR-21 used for ELINT and SIGINT missions.
(VI Brigada Aérea)

28

Mirage IIIEA I-006 was specially painted to mark the 25th anniversary of the type in Argentina.
(VI Brigada Aérea)

together with five F-16Cs and one F-16D from the 160th Fighter Squadron of the 187th Fighter Wing, Alabama Air National Guard. The US and Argentine aircraft undertook aerial combat training, this being viewed as a first step towards eventual Argentine participation in the Red Flag exercise, although this was never realised. Meanwhile, the Alabama ANG F-16s demonstrated their superiority over the Argentine aircraft, once again signalling that a replacement was due.

Yet another upgrade project was developed, this envisaging installation of a new radar, similar to the AN/APG-66 (a smaller version of this is used on the A-4AR), new weapons systems and electronics and a refuelling probe for a total of 39 million US Dollars. This upgrade never took place.

In 2001 an offer was considered to buy a number of ex-Qatar Air Force Mirage F.1EDA/DDA aircraft then in use with Ala 14 of the Spanish Ejército del Aire at Albacete. The aircraft were inspected by FAA personnel, but were not purchased. Also inspected

Flying low over the Pampas.
(VI Brigada Aérea)

Latin American Mirages

Mirages conduct a formation take-off from Mendoza during a deployment in 1998. (Santiago Rivas)

were F-16A/Bs in storage at the AMARC facility at Davis-Monthan in Arizona, with a plan to upgrade these to a standard similar to the Block 30. The latter would have provided a definitive replacement for the Mirage IIIEA, but again, budget cuts made acquisition impossible.

Exercise Águila II took place between 18-28 April 2001, involving 4 Mirage IIIEAs, 4 Fingers, 2 Mirage IIIDAs, 2 Mirage 5A Maras and 14 OA/A-4ARs, plus 7 F-16Cs and 1 F-16D from the 121st Fighter Squadron of the 103rd Fighter Wing, Columbia District Air National Guard. The next multinational exercise in which the Mirage IIIEA/DA took part was Ceibo 2005, held by IV Brigada Aérea at Mendoza, and featuring participation by the air forces of Argentina, Brazil, Chile and Uruguay.

Throughout their career, the Mirage IIIs were used to escort foreign aircraft visiting Argentina, including those carrying the leaders of Brazil, Cape Verde, France, Germany, Italy, Peru and others. Two Mirage IIIEAs and two Mirage IIIDAs also escorted the Aérospatiale/BAC Concorde supersonic airliner on its second visit to the country.

Since a replacement had been ruled out for the time being, it was again decided to improve the capabilities of the Mirage. I-017 was tested with a modification that allowed carriage of the AIM-9L/M Sidewinder AAM, as used by the A-4AR. In addition,

Two-seater Mirage IIIDA I-002 at the ramp of VI Brigada Aérea, at Tandil in 1999. (Christian Amado)

A Mirage armed with one R.530 and two R.550 missiles. (Juan Carlos Cicalesi)

new VHF equipment was installed and the Cyrano II radars were repaired. Meanwhile, a new upgrade plan had been developed by 2005, this foreseeing the installation of a new multimode radar, liquid-crystal displays, an inertial navigation system with a HUD, HOTAS controls and mission computer. All avionics were to be compatible with those of the A-4AR. The plan was to keep the Mirages viable until a replacement could be inducted, but again the government refused to approve the requisite budget.

Today, Argentina is the final operator of the Mirage III, together with Pakistan, and in 2006 the possibility of an upgrade was finally discarded for good. Although a new offer for ex-Royal Jordanian Air Force Mirage F.1s was received in 2009, the FAA wants to use the funds to purchase a batch of ex-Armée de l'Air Mirage 2000C/Bs to replace their older deltas. In the meantime, Argentina's Mirage IIIs continue to serve, 38 years after their arrival.

A Mirage taxies at Reconquista. Note the badge on the tail commemorating the 30th anniversary of the type in service. (Santiago Rivas)

Latin American Mirages

A spectacular dogfight display by two FAA Mirage IIIEAs, including I-018, during the Argentine Air Fest at Morón in 2010.
(Mariano Salcedo)

Argentina

Dagger and Finger

The Nesher/Dagger story

In the mid-1960s, in response to an Israeli request, Avions Marcel Dassault developed a simplified fighter-bomber variant of the original Mirage IIIC, the Mirage 5J. A French embargo on arms exports to Israel imposed in June 1967 prevented the delivery of the 50 Mirage 5Js, and the first 30 of these aircraft were instead adopted by the French Air Force (the other 20 were not built). Accordingly, Paris returned Israel's advance payments for the Mirage 5Js.

The French embargo on arms exports to Israel actually underwent two distinct phases. The initial embargo precluded deliveries of Mirage 5Js and their licence production in Israel, but was subsequently relaxed sufficiently for the IDF/AF to acquire huge quantities of spares for the Mirage IIICJs already in service. The second French embargo, introduced in 1969, was supposed to be 'total' – officially, at least. Unofficially, after only a short break, France re-launched spares delivery and the transfer of technology and know-how, probably in reaction to a major Libyan order for 110 Mirage III/5s, issued in late 1969.

In the meantime, a major project had begun in Israel to establish Israel Aircraft Industries (IAI) and expand the Bet-Semesh engine manufacturing company to enable the production of jet engines. These efforts received extensive assistance from Rockwell Corporation in the US, which already had facilities manufacturing automotive parts in the country. In 1970, Rockwell sent a number of highly experienced aircraft designers to Israel, among them Gene Salvay, who had previously worked for North American on designs including the B-45 Tornado, F-86 Sabre, F-100 Super Sabre and others. Salvay arrived in Israel in May 1970, just in time to witness the clandestine delivery of the first Mirage 5Js from a newly built batch of 51 aircraft of French manufacture (at least 8, and perhaps 10 two-seaters were to follow by February 1974). With extensive US support, the French had decided to deliver Mirage 5s in the form of knockdown kits – despite the arms embargo. Once in Israel, the aircraft were assembled by IAI, under close supervision from Rockwell engineers, who also helped install Atar 09C engines into a number of IDF/AF Mirage IIICJ interceptors.

The first Mirage 5Js entered service with the IDF/AF under the local designation Ra'am M in October 1971. Although all wore the manufacturer's plates of the French company Aérospatiale, they were officially heralded as the 'first fighter jets manufactured in Israel'. The provided explanation implied that Israel had acquired the necessary technical documentation via espionage, aided by sympathisers abroad, and foremost in Switzerland. This 'new' type was re-designated as the Nesher in autumn 1971 and saw extensive combat service during the 1973 Yom Kippur War. During the conflict and its immediate aftermath, IDF/AF Nesher pilots claimed in excess of 100 victories against various types flown by Arab air forces.[1]

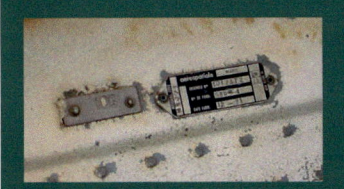

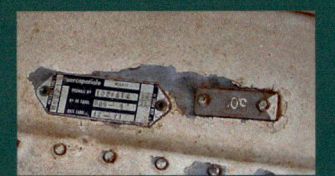

Manufacturer's plates on one of the Neshers/Daggers delivered to Argentina (Israeli AF serial number 501) show that the aircraft was manufactured by the French company Aérospatiale, and not by Israel Aircraft Industries.
(David Lednicer)

[1] Based on the article 'The Designer of the B-1 Bomber's Airframe', by Joe Mizrahi, published in *Wings* magazine, Volume 30, No. 4, August 2000

Latin American Mirages

With a requirement to acquire additional aircraft in the Mirage IIIEA/DA class, but for attack missions, the FAA began negotiations with the Heyl Ha'Avir by the mid-1970s, with a view to buying a batch of new IAI Kfir C2s. In the event, a US embargo on the Kfir's engines led to cancellation of the project. Instead, Israel offered a batch of 26 IAI Nesher jets, at a low price. The Nesher was a French-built Mirage 5, clandestinely delivered by France to Israel in 1970–71. The Nesher was equipped with the Snecma Atar 9C engine, also built in Israel by Bet Semes, and the aircraft was used extensively in combat during the Yom Kippur War, claiming numerous victories in air-to-air combat.

On 10 August 1978 the FAA signed a contract for the purchase of 24 single-seat Neshers (C-401 to C-424) for 4 million US Dollars each, and 2 two-seaters (C-425 and C-426) for 7 million US Dollars each. Delivery was to take place over a period of six months, and some days after contract signature the Argentine personnel departed for Israel to receive training. Another group went to Peru to train with Escuadrón de Caza 611 at Base Aérea Capitán Quiñónez González de Chiclayo, using their Mirage 5Ps.

On 9 November, Major Juan Sapolski was the first Argentine to perform a solo flight on a Mirage 5 in Israel, followed four days later by 1st Lieutenant Mir González. In October Captain Mario Pergolini and 1st Lieutenant Arnau joined Sapolski and González for air-to-air combat training at Eitam air base. Here they used the Argentine aircraft and learned the tactics successfully employed in combat by Israeli pilots. Other pilots followed them, including captains Puga, Kajihara, Donadille and Martínez and 1st Lieutenants Almoño, Dimeglio, Janett and Musso.

The aircraft received a 600-hour inspection, and the HF equipment was replaced by two VHF radios and an ADF. The ultimate intention was to equip the aircraft to Kfir C2 standard, but modifications would take place in Argentina since there was no time to complete them before delivery. A group of mechanics was trained at IAI facilities in Ben Gurion over a three-month period. Together with the aircraft, a batch of 50 Rafael Shafrir II infra-red AAMs was delivered. These were the first short-range missiles to equip the FAA, which at that time only operated the Matra R.530 on the Mirage III.

During delivery the Daggers were painted in an Argentine colour scheme but wore Israeli roundels.
(IAI)

34

The aircraft were delivered in a hurry, urgency being created by the imminent threat of war with Chile over the Beagle Channel. The aircraft, renamed as Mirage 5A Daggers, were embarked on the cargo ship *Jasper* on 26 November and sailed to Buenos Aires, from where they were taken to Aeroparque Jorge Newbery domestic airport. Here, the protection applied for the maritime trip was removed, and the aircraft were tested. Once they had been cleared for flight, on 12 December they flew to VIII Brigada Aérea at José C. Paz, joining the Mirage IIIEA/DA fleet. With the conflict with Chile resolved by the end of December, the aircraft were taken to Área de Material Río IV (ARMACUAR) at Córdoba Province, from 30 January 1979, for a major overhaul.

During January, when the new FAA commander in chief, Brigadier General Grafigna, assumed command, the Daggers made their first public appearance, flying over the Escuela de Aviación Militar (Military Aviation School), together with other aircraft.

On 16 August 1979, Base Aérea Militar (BAM) Tandil, 350km (217 miles) to the south of Buenos Aires, became VI Brigada Aérea, which would be the parent unit for the Daggers. On the same day the Escuadrón Mantenimiento Dagger (Maintenance Squadron Dagger) was created from technical personnel from Grupo 8 de Caza of the VIII Brigada Aérea. The aircraft arrived at Tandil from 28 August, and began their operational service at the base from which they would operate for their entire career.

On 26 November the first accident occurred, when C-406 suffered an engine failure close to the town of Azul, Buenos Aires province. The pilot, Captain Cimatti, ejected safely.

Fifteen days later the Escuadrón Dagger, created when the aircraft arrived, became Grupo 6 de Caza of VI Brigada Aérea, subordinated to the Comando de Defensa. By then, Israeli instructor Shlomo Erez, a veteran of Middle East wars, had arrived in Argentina to teach close air combat skills. He recommended different ways to use the aircraft, arming them with two Shafrir II missiles, two 1,300-litre (286-Imp gal) fuel tanks and two Rafael 30mm guns (similar to the DEFA 552) for combat air patrol missions. The Daggers differed from the Mirage III in having a maximum payload of 4,000kg (8,818lb) on seven pylons, two more hardpoints than on the Mirage. The fuel transference was also different, starting with the upper tank behind the cockpit.

According to the pilots, the Dagger was more manoeuvrable and docile during air combat. It was also unstable at lower speeds, which could lead to a loss of control, and with the 500-litre (110-Imp gal) upper tank empty, the jet could enter a flat corkscrew. In common with the Mirage IIIC that would arrive in the country later, the Dagger suffered from the problem of longitudinal instability, or porpoising.

In 1979 Grupo 6 de Caza was assigned 2,400 flying hours and recorded 2,385.5, the unit also employing an FMA/Cessna 182J for liaison. On 23 December 1980, the last aircraft to be delivered, C-419, was received after being overhauled by ARMACUAR.

In service

On 18 April 1980 the Daggers took part in the Aries gunnery training exercise from Neuquén city airport, in southwest Argentina. This was followed on 28 May by the Centauro exercise at IX Brigada Aérea, Comodoro Rivadavia, on the Patagonian coast.

Meanwhile, the first course for new Dagger pilots took place and on 2 July the initial pilots trained in Argentina performed their first solo flights.

Mayor Zabala, Mayor Villar and Captain Gómez, with two IAI assistants and Nesher 599 during SINT project testing in Israel. The aircraft later became C-427.
(Author's archive)

Exercises continued through the year, with Operativo Golondrina II at ARMACUAR and V Brigada Aérea, base for the A-4B Skyhawk. On 1 October the Daggers deployed to BAM Río Gallegos, on the southern tip of continental Argentina. Soon after that, on 7 October 1980, the two-seater C-425 was lost close to Tandil air base.

During the year, the brigade received Douglas C-47 T-18 and Aerocommander AC-500U T-134, while the unit as a whole flew 2,900 hours.

Since the FAA still intended to upgrade the aircraft, in 1980 development began of Project SINT (Sistema Integrado de Navegación y Tiro, Gunnery and Navigation Integrated System), which consisted of the addition of a Ferranti electronic unit (of which 11 were purchased), a Canadian Marconi HUD (two HUDs and a simulator were purchased), a British-built Doppler (30 purchased), an Elta EL/M-2001B ranging radar and an Elta Air Data Computer (ADC). With this equipment it was expected to attain a similar capability to that of the Kfir C2, and related work began in Israel in 1981.

In order to continue to augment the combat fleet, the FAA accepted an offer from the Heyl Ha'Avir for the air force's remaining Neshers, with the exception of one example that would be preserved in the museum at Hatzerim. On 22 September, the Dagger 2 contract was signed for 11 single-seaters, for 4.35 million US Dollars each, and 2 two-seaters, for 7.5 million US Dollars each. The single-seat aircraft received serial numbers from C-427 to C-437, while the two-seaters became C-438 and C-439.

On 29 May 1981 the second batch of Daggers arrived at Buenos Aires onboard the ELMA (Empresa Líneas Marítimas Argentinas) cargo ship *General San Martín*. The exception was C-427, which remained in Israel. The aircraft were taken to Aeroparque Jorge Newbery and in early June C-428, C-429, C-430, C-431, C-433 and C-438 flew to Tandil. The remainder followed in October. With their arrival at Tandil, Escuadrón 2 of Grupo 6 de Caza was created.

C-427 was used to test the SINT systems and received an air data recorder. In December the aircraft began flight tests, initiating the second phase of the SINT project.

Meanwhile, C-408 was being modified at ARMACUAR to serve as a prototype for the project and received the HUD, electronic unit, Doppler, ADC and radar. The servo-commands were modified to receive data from the ADC and a Thompson-CSF inertial navigation system (INS) was installed. Vicecomodoro Villar, Captain Rhode and Lieutenant Ardiles were selected to perform flight tests.

C-427 returned to Argentina at the beginning of 1982 and was enlisted by the FAA on 15 February. Soon, however, all work on the project was stopped with the outbreak of the Falklands War.

In combat

Once the news of the Argentine landing on the islands was known, on April 3 the order was given to Grupo 6 de Caza to prepare two squadrons, each of six Daggers, to be deployed to the south. A third squadron (Escuadrón 1) was to remain at Tandil.

VI Brigada Aérea had 34 single-seat Daggers and 3 two-seaters in its inventory. The last three were all in service, while of the single-seaters, C-402, C-405, C-422, C-423 and C-424 were being overhauled and did not fly before the end of the war. Furthermore, C-408 and C-427 were at the Area de Material Río IV for the SINT program.

Many of the other Daggers were not operational when the war began, but they were put into service and deployed before the end of the conflict. Of the aircraft in service, only C-413 and the two-seaters were not used on combat missions.

Escuadrón III, later named 'Las avutardas salvajes', was established from the basis of Escuadrón III of Grupo 6 de Caza, under the command of Major Carlos Napoleón Martínez and received the order to deploy to BAM Gallegos, at Río Gallegos. The nickname 'Las avutardas salvajes' (Wild Bustards) referred to a bird found in the area around Base Aeronaval Almirante Quijada at Río Grande.

Escuadrón II Aeromóvil, later called 'La Marinete', was created from the basis of Escuadrón II, under the command of Major Juan Sapolski, and was sent to IX Brigada

A Dagger with 14 250lb bombs built in Israel.
(FAA)

Aérea, at Comodoro Rivadavia. The unit's nickname, 'La Marinete' was derived from a song composed by 1st Lieutenant Mario Callejo, entitled 'La Marinete del Escuadrón Dagger'.

On 5 April, two C-130HJs and one Boeing 707 began to carry the necessary support equipment to the south, while a Fokker F28 took the personnel who would prepare the aircraft for operations from their deployment bases.

On the following day the 12 Daggers departed to the south. The first formation took off at 11.00 with 1st Lieutenants Luna (C-401) and Senn (C-416), Captains Norberto Dimeglio (C-420), Díaz (C-429), Horacio Mir González (C-421) and Moreno (C-436), with Major Sapolski (C-431) as leader. Major Martínez (C-410), Captains Janett (C-433), Robles (C-428) and Cimatti (C-430), and 1st Lieutenant Ardiles (C-412) followed them at 16.30. Aircraft C-401, C-410, C-412, C-428 and C-430 would be sent to Río Gallegos, while the others were intended to go to Comodoro Rivadavia. Despite this, when they reached their destination, C-401, C-412, C-416 and C-420 were unserviceable, so the decision was taken to send the remaining eight jets to Río Gallegos and keep the others there until more aircraft arrived from Tandil.

Because BAM Gallegos already supported a squadron of A-4Bs, it was decided to send Escuadrón I Aeromóvil to Base Aeronaval Almirante Quijada at Río Grande, Tierra del Fuego, where the Dassault Super Etendards of the Comando de Aviación Naval were operating. Eight Daggers flew to the base on 7 April, while on 13 April Escuadrón II Aeromóvil was reinforced with the deployment of C-407 and C-434 to Comodoro Rivadavia.

On the following day, and in order to familiarise themselves with the zone in which they would be operating, Captain Mir González (C-436) and Lieutenant Volponi (C-421) made a flight to the islands from Río Grande. Meanwhile, on 16 April two aircraft made a similar flight from Comodoro Rivadavia, with Major Sapolski (C-407) and Captain Díaz (C-434) at the controls. In both cases the aircraft flew with full ammunition for their guns and three 1,300-litre fuel tanks. Because of technical problems C-420 was sent to Tandil for repairs and was replaced by C-404 on 20 April. On the flight made to the islands by the aircraft of Escuadrón II, it was clear that IX Brigada Aérea was located so far away (720km; 447 miles), that it was impossible for the aircraft to complete the long flight with a full load of weapons and fuel. It was therefore decided to deploy the aircraft to the airport of the city of San Julián, in Santa Cruz province. BAM San Julián was created at San Julián, and the A-4Cs of Grupo 4 de Caza would also operate from here. In order to operate the Daggers from San Julián it was necessary to install aluminium plates at one end of the runway, creating a small apron for the fighters. In addition, a taxiway was created to connect the runway with Route 3, allowing it to be used as a dispersal in case of air attack.

During the war the Daggers employed two weapons configurations. For air-to-air combat this comprised guns and two Rafael Shafrir II missiles. For air-to-surface attack, the weapons consisted of guns and two or four Spanish-made Expal (Explosivos Alaveses) bombs of 250kg (551lb), and three 1,300-litre fuel tanks. The bombs were fitted with conventional tails (BR) or with parachutes (BRP, Bomba Retardada por Paracaídas, parachute-retarded bomb). The aircraft based at Río Grande also employed a configuration with two 1,300-litre tanks and a single Mk 17 1,000lb (454kg) bomb. The British-built Mk 17 was also used by British Sea Harriers and Harriers during the war and by Argentine Skyhawks and Canberras. Later in the war, the Río Grande air-

Daggers head out for an attack mission against British forces, guided by a Learjet of the FAA. The fighters carry bombs on the rear fuselage pylons. (FAA)

craft were also modified to carry two of the 1,700-litre (374-Imp gal) tanks used by the Mirage IIIEA. For ferry flights the Daggers carried two 500-litre (110-Imp gal) tanks underwing and one ventral tank of 880-litre (194-Imp gal) capacity.

Escuadrón I Aeromóvil

This unit remained at Tandil, using the aircraft that were not deployed south. Most of the aircraft were not operational at the beginning of the war but were later put into service. The only single-seater not to be sent south was C-413. Together with the three two-seaters (C-426, C-438 and C-439), this was used to qualify new pilots, conduct training and provide air defence for the Buenos Aires province. The squadron maintained a constant quick reaction alert force, with two aircraft at the end of the runway, armed with two guns and two Shafrir missiles. The two-seaters were sometimes also used to carry spares to the south, flying without guns and using this space to carry small loads. During April and May, 1st Lieutenants Musso, Dellepiane, Demierre, Piuma Justo, Gabari Zocco, Ratti, Reta and Valente were qualified and all of these, with the exception of Reta, were sent south to take part in the fighting. Alferez Peluffo and Penacchio were also trained on the Dagger but were not deployed.

On 4 June, the squadron was surprised when a formation of 10 Mirage 5Ps of the Fuerza Aérea del Perú (FAP, Peruvian Air Force) requested permission to land. The FAA had purchased the aircraft the previous year, but their delivery was only planned to take place much later. The 10 aircraft had been prepared quickly and were flown to Argentina by Peruvian pilots, to replace war losses. They were immediately pressed into service with the FAA. Argentine roundels were painted over the Peruvian markings and the aircraft received the serial numbers of Daggers lost in combat, becoming C-403, C-404, C-407, C-409, C-410, C-419, C-428, C-430, C-433 and C-436.

All these Mirage 5s had been delivered to Peru between 1968-74 and were sold to Argentina by the end of 1981, with Peru expecting to replace them with Mirage 2000Ps. On 14 June the first flights were conducted by C-404, C-409 and C-410, but the end of the conflict prevented their deployment.

Latin American Mirages

Escuadrón II Aeromóvil 'La Marinete'

On April 25, San Julián saw the arrival of C-401 (1st Lieutenant Román), C-407 (Captain Dimeglio), C-412 (Major Juan Carlos Sapolski), C-416 (1st Lieutenant Mario Callejo) and C-434 (Captain Díaz). These were followed by 1st Lieutenant Senn in C-432 one day later, and by 1st Lieutenant Ratti in C-421 on 28 April. C-432 was sent to replace C-401, and C-421 to replace C-416. Both had to land at Tandil because of fuel leaks, but the aircraft were repaired and returned to San Julián shortly after.

The first combat order arrived on 27 April and called for the preparation of C-407, C-412, C-416 and C-432 with two BRP bombs and three 1,300-litre tanks each. Their mission was to attack the British fleet approaching the islands. The jets were to be covered by C-404 and C-434 armed with two Shafrir II missiles. However, the ships were not found, and the mission was cancelled.

On the following day, yellow stripes were painted on the wings and tails of the Daggers for easier recognition by Argentine troops.

On 29 April, a new order to attack the British fleet arrived, but again the mission was cancelled. At 18.30 an alert arrived concerning a possible raid on the airport, and the order was given to send the Daggers to Tandil, where they arrived between 22.00 and 24.00. The exception was C-412, which remained at San Julián. On the next day C-404 returned, but the others were grounded by bad weather. C-401, C-403, C-407, C-421 and C-432 returned at 11.00 on 1 May.

On their return, the Daggers were immediately prepared for combat. OF.1101 had arrived, ordering the departure of Fierro flight on a CAP mission, armed with guns and Shafrir missiles. Fierro flight consisted of C-421 flown by Captain Raúl Díaz and Lieutenant Aguirre Faget in C-412. The second aircraft remained on the base after suffering engine problems. Díaz recalls the mission. *'Close to the islands I made contact with the Centro de Información y Control (CIC) of Puerto Argentino, who directed me to an intruder that was at 60 miles (97km) to the east of the islands.*

'I armed the missile and gun panel, selected bearing 090 degrees and followed the instructions of the CIC. They informed me the intruder was 6,000ft (1,829m) below. The controller continued indicating course and distance, and we were approaching from the front. When 12 miles (19km) separated us, the intruder descended fast and the radar lost him. This led me to think that the enemy was refusing to engage in

Daggers stand ready for their next mission at the end of San Julian's runway.
(FAA)

combat. The radar operator made me return to the west, as I was some miles to the east of the islands. There was a cloud layer that made it difficult to see the enemy ships. Then the controller said I was flying over a radar echo, which could be a warship. The controller told me to take care as the ship could attack with missiles, but this didn't happen. When I was again over the islands, the operator told me about an air strike made by the Sea Harriers over BAM Malvinas.

'*I asked them if they wanted me to go there, but they answered no, because the aircraft were inside the AAA defence sector. Then he alerted me about the presence of an echo flying towards me from the southeast, very fast and climbing. It was detected at a distance of 18 miles (29km) from me and I turned south on a collision course and kept a height of 26,000ft (7,925m), accelerating until reaching 450kt. I saw the instruments before we crossed but I had fuel for no more than five minutes before reaching the minimum required for the return.*

'*When I was only 8 miles (13km) from the intruders and 3,000ft (914m) above them, they decided to abandon the engagement. They descended violently until our radar lost contact. I tried to see them without losing height, but I didn't make it. Then I heard the voice of the controller asking me how much time I had until I had to return, and I replied that it was time to return to my base.*'

The next mission marked the first attack sortie by the unit, and the first attack pressed home by any Argentine aircraft against the British. Using the Torno callsign (OF.1105) three Daggers were sent to attack the destroyer HMS *Glamorgan* and the Type 21 frigates HMS *Arrow* and *Alacrity*, which were bombarding Argentine positions near Puerto Argentino/Stanley. The flight was commanded by Captain Norberto Dimeglio (C-432), with Lieutenant Gustavo Aguirre Faget (C-412) and 1st Lieutenant César Román (C-407). The pair took off at 15.45 armed with three BRP bombs. Román remembers that '*the target was indicated as being to the north of Puerto Argentino, about 15 miles (24km) off the coast. We flew to the north of Gran Malvina/West Falkland and then Soledad Island/East Falkland. Close to the target I saw a helicopter to our left. We were at four minutes.' I told the leader and he said 'We will continue to the target! Then we saw something on the horizon and we began the attack – it was a rock. We changed course, still following the coastline. We arrived in the target zone but there was nothing, so we continued to Puerto Argentino. We started to see that someone was firing against the coast and in front of us we saw some explosions like fireworks. Then we saw them: there were three frigates very close to Puerto Argentino, constantly firing.*

'*I heard on the radio the leader saying 'Number 1 to the ship in the middle, number 2 to the one on the left, and number 3 to the right!' We flew as low as we could, over a calm and grey sea, with low clouds that were at no more than 1,000ft (305m).*

'*We pulled to full throttle for the attack. I saw hits on the water and I thought they were already firing against us, but they were the guns of the leader. The surprise was total, because it was the first attack against the British fleet. I didn't fire with my guns, as I wasn't convinced the ships were theirs and I believed they could be from our Navy.*' The leader's bomb was dropped. The number 2 later said that he climbed to attack at an angle and they fired against him with everything they had. From the coast, Major Catalá saw when the ship launched two missiles against them, which passed below the aircraft.

'I saw a big explosion on the water. We flew through the clouds and we lost each other climbing. We were returning individually. Then we heard the radar operator saying 'The Tornos have bandits on their tail!'.'

Aguirre Faget also remembers that *'from San Carlos Strait (Falkland Sound) we flew at very low level, at 420kt, then we accelerated to 480 and I think that on the last moment of the attack we were at 520. We tried not to pass 540kt due to a limitation on the fuses. The three of us tried to keep the formation, at about 50m (164ft) on a lateral line, something difficult at very low level. In those times we didn't have a radio altimeter and we calculated height visually – about 50ft (15m), but when we accelerated on the final part of the attack and we had the ships in front of us, it's possible we had climbed a bit and the formation was lost.*

'We had two VHF radios. The pilot could select the green button, the red or both; we had one frequency to talk to the flight and on the other we had three different frequencies: one aircraft had the Malvinas radar, the other the BAM Malvinas tower and the other the relay aircraft, which was BAe 125 serial LV-ALW of the Fenix Squadron, flying at 40,000ft (12,192m). When we were flying to the north of Isla Soledad (East Falkland) I heard on the radio 'Don't shoot, stop!' Later I knew this was Captain Gonzalez, who was acting as an observer for the AAA close to the airport. He saw the Mirage IIIEA of Garcia Cuerva approaching to land at the airport and the artillery was firing on to him. When I heard that I looked south and at great distance I saw the lines of the tracers, going into the clouds. I heard that for about five seconds. Later I knew that it was the moment when Garcia Cuerva was shot down'. Aguirre Faget was sure that the pilot had made a 'general ejection' – releasing all the external stores, including the pylons and the missiles, which fell inert. *'Since AAA gunners and troops on the ground had been killed and wounded by British bombing, when they saw the aircraft release its load, they opened fire. We knew we had to be alert with the batteries at Goose Green and Puerto Argentino (Stanley), as we were forbidden to fly over there, except with special coordination.*

'We came in from north to south, and we passed Bahía de la Anunciación (Berkeley Sound). I supposed we would find the ships there and we had a little fuel. Dimeglio said 'We will continue for another two minutes', because we flew over the target zone and we found nothing.

'When we turned, passing Bahia de la Anunciación, close to the coast was a helicopter, like a Sea King. It wasn't a Lynx; it was a big helicopter, possibly directing the fire. Roman almost crashed into it on his turn. We passed the helicopter and then we saw the ships. We were about 5 or 6km (3 to 4 miles) from the coast' – there were a lot of witnesses on the coast, who saw the aircraft very small but they saw everything that happened.

'Besides the ship I attacked, closer to the coast was another one and behind my ship was another showing her side. Roman told me that because they hadn't fired on him he had some seconds to prepare his attack but was doubtful as he was not sure that the ships were not Argentine'. Aguirre Faget then selected the target he saw to the east. *'I was sure they were British. When I had her 4,000 or 5,000m (4,374 or 5,468 yards) ahead of me I didn't have her on my sight. The sight was on the water, so I climbed to about 3,000ft (914m), then lowered the nose with negative g. I waited to have her inside the range of my guns and fired. Close to the stern I levelled, dropped the bombs and escaped to the east.'*

The aluminium-plate apron hastily laid at the end of the San Julian runway. (Oscar Arredondo)

Immediately, the ships' port and starboard Oerlikon 20mm guns opened fire, as did the Bren machine-guns on the bridge. *'The artillery fire looked like arrows with smoke and it seemed as if all of them would hit me'*, Aguirre Faget recalled. To the Argentine pilot, it seemed as if the shells were going to hit his aircraft but shortly before, the tracers seemed to move to the sides, and the aircraft escaped the AAA. *'I fired at the origin of those 'arrows'. I probably started to fire before I was within range, because when they are firing at you it makes you fire sooner. Then I levelled, dropped the bombs and turned to the east. I saw nobody, neither Dimeglio nor Roman. The ships were in a triangle and on my flight path I attacked from stern to bow.*

'As I approached they opened up with a very concentrated fire – that's why I cannot say which ship I bombed. I was concentrating on my aircraft, the sight, the speed, as if I was on a bombing exercise. I tried to keep the performance to allow my bombs to hit the ship. It's not courage or cowardice, you are there and you have to do that, the best I could do was to keep firing with my 30mm guns and the bombs'. Aguirre Faget opened fire and since he was now flying too high, descended down towards the water. *'After dropping the bombs and while I was descending I began porpoising, which I think saved me. As I looked to the ship they were still firing at me – I also saw smoke. I don't know for how long, but I think for two or three minutes I flew to the east to escape from the gunfire'*. According to witnesses on the coast, a missile was launched against Aguirre Faget as he climbed to attack. Since the missile passed below his jet, he did not see it.

Writing in his logbook, Ian Inskip, the navigating officer on HMS *Glamorgan* described the following events at 19.45 GMT: *'Flight Deck Officer jumping up and down on Helicopter Intercom – three aircraft closing from right astern. They rounded Cape Pembroke Point at deck level, and came at us all taps open. The PWO ordered 'Come very hard left' to open weapon arcs. I made a quick check on the chart, vetoed the order, and said come very hard right, as we were only a couple of cables from the minefield. The aircraft was indicated to the port Seacat [surface-to-air missile], but as we turned the other way, port Seacat lost it in the blind arc and starboard Seacat did not have time to acquire.'*

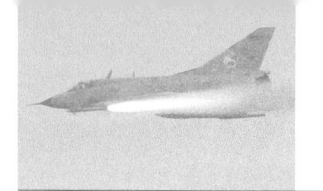

Aguirre Faget continues. *'After the attack, about five or six minutes later, Dimeglio said 'One' and I had to say 'Two' and Roman 'Three' to know that we had all survived the attack. But we said nothing and he shouted 'Answer, assholes!', then I said 'Two and Roman 'Three'. To return to San Julian meant flying over the artillery of Puerto Argentino and we had limitations because of the artillery and our own missiles. To gain height I had to fly at about 300kt, and I expected to level the aircraft initially at 20,000 or 23,000ft (6,096 or 7,010m) and then continue climbing. At speed, the Dagger is a marvellous aircraft, but it takes time to accelerate. When I was climbing, the Malvinas radar informed me that our number 2 had two Sea Harriers following him. I realised I was slow. I could eject three extra tanks but I only ejected one. I tried to return with the other two tanks because I knew we only had a few. I had to do a very precise navigation to return with enough fuel. I couldn't level my aircraft or change course. I had to reach cruise level and then the aircraft would start to accelerate.*

'When we finished the attack we could put the transponder on active or standby mode, and when I had the British aircraft behind I turned on to confirm if they were pursuing me, and it confirmed that. With the sun in front of me I saw nothing. It was 17.00, I had no ammunition and little fuel.

'The pursuit continued while I was climbing, first at 12 miles (19km), then at 9 miles (14km), and when they told me the aircraft were at 6 miles (10km) I started to breathe again as I was accelerating my jet to Mach 0.9.'

The attack mission was covered by Fortin flight (OF.1107), with Captain Guillermo Donadille in Dagger C-403 and 1st Lieutenant Jorge Senn in C-421. Fortin flight was orbiting over Gran Malvina/West Falkland at 30,000ft (9,144m). Because of the Sea Harriers' pursuit against Aguirre Faget, Fortin flight was sent by the ground controller to protect him. They put their aircraft behind the British jets, at a distance of two miles, but their Shafrir 2 missiles locked on to the sun and they could not fire. Donadille's aircraft suffered an electrical failure that meant he could not fire his guns. Over San Carlos, both Sea Harriers began a descending turn and disappeared from the radar.

Aguirre Faget returned to base almost without fuel. *'On the Dagger, if you made a bad approach for landing you could not accommodate the aircraft on the runway, so you had to make a turnaround, consuming 150 litres (33 Imp gal) of fuel. When I was about to land I entered very large clouds. When I was approaching I started to search for VOR radials for other bases, to try to see the distance to San Julian. On the route, close to the coast, I saw an aircraft, a contrail, flying parallel to me. I didn't know if it was a Sea Harrier but I suspected it was one of my wingmen. I kept radio silence, watching the engine functions and calculating when I had to start the descent – if I calculated it wrong I would consume more fuel and maybe have to eject over the sea. The concentration lasted until I shut down the engine. On the ground I scolded the mechanics because I didn't believe them when they said the aircraft had received no impacts, but in fact it didn't.'*

Because of a failure on the photometer, the film from the Omera 110 camera was blurred. *'I couldn't believe it, because I had seen which ship it was, my angle of attack, which ammunition they were firing, and when I was pressing the bomb re-*

Daggers at San Julián. (FAA)

lease button and making the escape. The photometer has a screw that moves to show if the light is right or not, and the petty officer in charge said to me after landing, 'If you saw it wasn't moving, you should have done something, because maybe it was locked'. But at that moment, in the middle of the attack... I wanted to kill him!'

The Daggers of Torno flight damaged *Glamorgan*. According to Ian Inskip, *'the middle aircraft came for us in a very shallow dive. He attempted to strafe us and cannon splashes made tracks towards the stern. Just before they reached the stern, the pilot pulled up, took his finger off the trigger. Then 'the wings fell off!' – or rather everything on them. There was no way the bombs were going to miss, until their parachutes deployed. By reversing our turn, we had spoiled his aim and the two 1,000lb bombs went in either side of the quarterdeck. There were two huge explosions that lifted the stern 17ft (5m). Columns of smoke were coming out of the after funnel as the gas turbines were being flashed up. The centre aircraft overflew us, afterburners on. One Dagger flew up the port side, and one up the starboard side, both below bridge level. The other two Daggers went for* Arrow *and* Alacrity. *All three ships opened up with their 4.5in (114mm) guns.* Arrow *was strafed in the funnel and* Alacrity *was splintered by a near miss.'*

HMS *Arrow* was attacked by Dimeglio and hit by 11 30mm rounds, while HMS *Alacrity* also received some minor damage from Roman's attack, although he had not fired his guns as he was not sure the ship was British. The ships then departed to the Carrier Battle Group. None of the Argentine aircraft was damaged, despite the intense AAA fire, and all landed safely between 18.25 and 18.40.

The Fortin flight was the last CAP sortie flown by this Dagger squadron during the war. From then on they would only be used for air-to-surface attack.

On 2 May, OF.1129 and 1130 were received, these outlining a mission using four Daggers armed with guns and two BRPs to attack naval or ground targets. However, because of bad weather, the British forces did not approach the islands on this day.

On the following day C-415 was deployed to the squadron and C-434 returned from Tandil from where it had departed on 29 April. In the days that followed the British did not approach the islands, and bad weather prevented any attack mission being flown. Meanwhile, C-403 flew with Captain Dellepiane to Río Gallegos to test the use of the 1,700-litre tanks used on the Mirage III. It was decided only to use these tanks in concert with the Daggers based at Río Grande, and to use the 1,300-litre tanks on the Daggers based at San Julián. C-403 returned to its base on 11 May.

On 6 May, OF.1166 and OF.1167 were released, ordering the preparation of the Fierro and Tucán flights, each aircraft armed with two 250kg (551lb) bombs, but again the operations were cancelled. Póker (OF.1186) and Coral (OF.1187) were also cancelled, each comprising three aircraft.

New orders arrived three days later, and Puma flight (OF.1175) took off at 13.00 with Major Sapolski in C-401, 1st Lieutenant Sinn in C-407, Captain Díaz in C-432 and 1st Lieutenant Vallejo in C-412. The aircraft approached the islands in bad weather and each pair lost sight of the other. Because of the weather and the fact they could not communicate with the CIC on the islands, they decided to return while they were close to Sebaldes/Jason Islands, landing at 15.40.

Meanwhile, at 15.00, Jaguar flight (OF.1176) took off with Vicecomodoro Villar (C-404), 1st Lieutenant Roman (C-420), Captain Dimeglio (C-434) and Lieutenant Aguirre Faget (C-415). The aircraft liaised with the CIC but the latter did not have the Daggers on radar in order to guide them to possible targets. Also, the bad weather meant the Daggers could not descend to low level. When they were above the islands, Villar ordered the return at 15.50, and they landed at 16.15.

Since the British ships could not be found in the days that followed, the Daggers' activity was limited to some training and test flights. On 11 May, Vicecomodoro Villar flew C-432 on an armed reconnaissance mission close to the base, while on 12 May C-421 was used for two training flights. Two days later, training flights were made by C-404, C-415 and C-420.

On 18 May, C-401 was sent to Río Grande, joining Escuadrón III until 24 May.

A further armed reconnaissance mission was ordered on 20 May, and three jets took off at 11.00. These comprised C-432 with Major Piuma, C-412 with Lieutenant Carlos Castillo and C-415 with Captain Dimeglio. The aircraft failed to find any target and returned to base, landing at 13.40. Later, at 16.00, C-404 with Captain Guillermo Donadille and C-421 with Major Puga took off. Again, they found nothing, and landed at their base at 17.15.

Over San Carlos

As soon as news of the British landings was received, the order was given to Argentine combat units to prepare aircraft to attack the ships. Very early on the morning of 21 May, OF.1183 and 1184 arrived over the islands, each package comprising three Daggers armed with guns, two BRPs and two 1,300-litre tanks. The first to take off was León flight, with Captain Norberto Dimeglio (C-404) and Lieutenant Carlos Castillo (C-407). Dimeglio recalls the mission: *'Shortly before 10.00 we started the engines. 1st Lieutenant Senn's engine didn't start, so he remained on alert for the missions in the afternoon. We flew side by side, and separated in order to search the sky for possible enemies. We arrived at the islands and began a very low flight, as close to the ground as we could. Now we had only 30m (98ft) separation between one aircraft and the other, at almost the speed of sound. After some minutes we arrived over the hills that are to the west of San Carlos Strait. After crossing them we lost their protection.*

'Then an unforgettable sight met us: on the waters of the sound were no less than eight British ships. One of them was not far from us (it was the 'County'-class destroyer HMS Antrim*). We were approaching diagonally from the stern. This meant*

Dagger C-420 at San Julián, with Peruvian external tanks. (FAA)

they didn't have time to use their gun on the bow and they also didn't have time to use their missiles.

'We broke radio silence and shouted words of encouragement for the attack. We went closer to the water, achieving total surprise. When we were within range for our guns we opened fire, aiming at the upper decks and the radar antennas. I saw a trail of yellow and red stars hitting the target and then crossing, creating a spider's web of lights. The moment to drop the bombs arrived. We launched them with their retarded fuses, flew over the ship, and with full power we turned tight to our right and headed for Gran Malvina/West Falkland.

'At that moment my aircraft started to vibrate slightly and when I looked at the speed indicator, I realised I was almost at the speed of sound. I called anxiously on the radio and I was happy to hear that we had both escaped without problems. When we were returning we passed the three aircraft of Perro flight (from Escuadrón III) flying to attack the same target, and we told them what was waiting for them.'

On *Antrim*, Chief Petty Officer Lionel N. Kurn, the ship's helicopter artificer, recounted the raid. 'We'd been told the Falkland Islands were too far away from the Argentine mainland for their air force to be a threat. But there they were – so it clearly wasn't true'. Chief Petty Officer David Heritier, Fleet Air Arm, was also aboard *Antrim*: 'About mid-morning, another Pucará came skipping across, rolling over the headland and disappearing very quickly. Then A-4 Skyhawks (actually the Daggers of Leon flight) were on us, coming straight down the sound. The ship was turning at full power. They came astern, going for us. I was standing by the door and the flight commander suddenly shoved me down into the fuel space where we made a big bundle on the floor.

'The two A-4s [sic] hammered over the top of the ship with a tremendous roar, leaving the back of the ship enveloped in steam. Our pyrotechnic locker, were we kept flares, marker marines, phosphorus and other nasty things, was pouring smoke. We had fire hoses rigged anyway, as we were at flying stations. We didn't know a bomb had gone into the ship (in fact, the only hit achieved was on the second attack by León flight), and assumed this was a fire below decks, so we commenced a boundary cooling to stop it spreading… The unexploded bomb was discovered in the bathroom just below our feet. The damage control party wedged it so it wouldn't roll around.'

These aircraft were followed immediately by Zorro flight with Captain Díaz (C-412), Lieutenant Aguirre Faget (C-415) and Captain Dellepiane (C-434). Arriving at San Carlos they saw the frigate HMS *Brilliant* and went to attack her. Because of a failure of the electrical systems, the bombs did not release and the Daggers could only attack with guns, making 16 hits on the ship. '*The surprise was total*', Díaz remembers. '*The British supposed that the meteorological situation that lasted until a few minutes before would last longer, reducing the activity of Argentine aircraft. Because of this they almost didn't have air cover. When we arrived over the sound, the view exceeded the expectations we had conceived in our minds. We saw clearly the defensive disposition of the British ships, with the warships on a semicircle around the landing zone. We flew at about 530kt against one of the closer frigates. We approached at very low altitude, in a line formation. From a distance of about 1,000m (1,093 yards) we started to attack a Type 22 frigate with guns. When we were at the correct distance to launch the bombs, a failure occurred in the launch system as a result of the extreme cold, and the bombs were not dropped. The frigate didn't have time to react to the attack; instead it tried to escape from our gunfire at maximum speed. Another ship was to our right and when they saw us they opened fire intensively against us, but it wasn't very effective.*'

During the attack C-434 received light gun damage on the left air intake, while C-412 took a hit on the VHF antenna. The HMS *Brilliant* received some damage to the operations room and some sailors were wounded. The damage put the Sea Wolf and Exocet missiles and the sonar systems out of operation.

More missions were ordered in the afternoon, completing the first day of operations over San Carlos by the Daggers from San Julián. The Fuerza Aérea Sur sent OF.1198 and 1199, the first of these comprising Ratón flight, with Captain Donadille (C-403), Major Piuma (C-404) and 1st Lieutenant Senn (C-407). OF.1199 was Laucha flight, with 1st Lieutenant César Román (C-421), Major Puga (C-412) and 1st Lieutenant Callejo (C-415). Both flights took off at 14.00, and Laucha was the first to approach the target zone.

Román describes the action that followed. '*The last stage of the flight was made bordering the northern coast of Gran Malvina/West Falkland. Once we passed Rosalie Mountain we found the waters of the sound and, after flying very low, we discovered that we were higher than expected. The eastern shore is high and abrupt, so we pushed the stick forward to scrape the waves and thus avoid the ship's radars. I saw a warship near San Carlos Water, with some flames and a long column of smoke pouring from it (HMS* Ardent*). With my number 3 (Major Puga), I attacked a frigate. The ship's AAA raised a wall of water and in fact I went through one of these geysers, which considerably reduced visibility and aiming. Together with number 3, I fired the cannon and dropped the bombs, and then we egressed at full speed on an easterly heading, while number 2 attacked another warship to the north of us*'. Román and Puga attacked the *Brilliant*, hitting the ship with their 30mm guns, and damaging its Lynx helicopter XZ732. Callejo attacked another frigate without hitting her. Aircraft C-412 and C-415 were hit on the windscreen and, after they had landed, were withdrawn from service.

Meanwhile, Ratón flight was preparing to depart. Piuma had trouble starting his engine, so the Lauchas took the lead after Donadille gave authorisation. Once the problem had been solved, they flew straight to the target area. Around 92 miles (148km) from the islands, the three Daggers were skimming the waves. Just before arriving at Gran Malvina/West Falkland they were informed by the air controller that there was considerable Harrier activity in San Carlos Strait and to the north of the sound. Over

Three Daggers at San Julián. (FAA)

Gran Malvina, the aircraft were now in a single line, side by side, separated by about 300m (984ft). Donadille was in the middle, Piuma to the left and Senn to his right.

Stationed in their patrol area under the control of HMS *Brilliant*, Lieutenant Commander Nigel Ward in Sea Harrier ZA175 and his wingman, Lieutenant Steve Thomas in Sea Harrier ZA190, were flying CAP over Gran Malvina.

Three minutes away from the target, with the sound already in sight, the Daggers were low over the ground and had accelerated to 500kt. Senn now warned his fellow pilots: *'Watch out, one aircraft to the right'*. Senn had seen Thomas passing from right to left, crossing above the Daggers. Donadille looked in the direction he had been warned and saw an aircraft slowly overtaking them at approximately 50m (164ft) altitude and 400m (1,312ft) to the right of Senn. At first, because of the profile and dark colour, Donadille thought the jet was an A-4B, although it did not make sense to him that it was flying faster than a Skyhawk. At about this moment, Piuma climbed with afterburners turning right (to the south), knowing that the height would give him an advantage during the combat. With altitude, he would be able to protect the tail of the other two Argentine jets.

As both Harriers, flying in their parallel circuits, turned north, Thomas looked below and behind him and saw two triangular shapes flying east at high speed: Donadille and Senn. Thomas only saw the Argentines when he was above them. He alerted Ward and immediately gave chase, turning his height into speed to try and catch the Argentine aircraft.

At that moment, the dark aircraft that Donadille saw (Ward) turned to the left and Donadille realised it was a Sea Harrier. The Ratones leader commanded: *'Drop ordnance and face him'*. Senn replied: *'He has not seen us'*, but Senn was referring to Thomas's Sea Harrier that had initially crossed above them. Senn did not realise that the aircraft that Donadille was looking at (Ward) was turning to get on his tail. Donadille commanded sharply: *'Drop the ordnance, damn it! And break right!'* Finally Senn did as he had been ordered and started to turn gently to his right.

Ward was turning in such a way that he was going to get on Senn's tail before Donadille could prevent it, so in despair he fired his guns in the general direction of Ward. Ward later said that he was surprised that their enemies were staying to fight instead of continuing their path to attack the Task Force, or even returning to their bases. At this moment, Donadille was 90 degrees to the side of Ward. The firing guns illuminated the belly of the Dagger and at that moment Ward broke to his left, now flying head on against the two Daggers. Donadille was higher and Senn lower.

Discovering that his Dagger had not been loaded with tracers that would have helped his aim, Donadille then dove towards Ward, who was approaching head on and diving. Donadille was firing again, this time ahead of the Sea Harrier. Focused on his aim, Ward flew 400m (1,312ft) away and below from Donadille, passing extremely low over the ground. Donadille started to climb and avoided the ground by less than 30ft (9m).

At this moment, unknown to Donadille, Thomas had managed to get on his tail. Piuma was still climbing and turning right and Senn was now turning to his right very sharply. When Donadille started to climb, he found Senn in his path. To avoid colliding with his wingman, who now had condensation trails on his wingtips, Donadille was forced to turn left (and put his aircraft almost inverted), crossing very close to his wingman.

A few seconds later, turning again to his right, and trying to see what had happened to the Sea Harrier he had fired at, Donadille received the impact of the first of Thomas's missiles. The damaged Dagger began to porpoise, and a few meters above the ground entered a flat spin. The control of his aircraft lost, Donadille ejected and no more than 20 seconds later, he touched the ground. Approximately 500m (1,640ft) ahead of him, his Dagger crashed and erupted in flames.

By now, Piuma had levelled his aircraft and was able to see the combat taking place below him, to his right. Two Sea Harriers were getting on Senn's tail. Senn had now reversed his turn, going to his left (west) and trying to face the threat on his tail. Ward entered a turn so tight that it surprised Piuma, still watching from above. Thomas was now approximately 700m (2,297ft) distant and below, at 90 degrees to the tail of the Dagger. The missile departed the left wing of Thomas's Sea Harrier and Piuma called to Senn to break his turn, but the Sidewinder approached too quickly. C-407 was hit above the left wing, close to the tail.

Piuma called Donadille on the radio, but received no answer. Then he saw the second Sea Harrier (Ward) turning gently to the left (north) below him and at around 120ft (37m) over the ground. Piuma dove on him. Ward and Thomas had started to fly together in their CAP, heading east. Piuma got inside Ward's turn and from a distance of 600m (1,967ft), 30 degrees to the left and slightly above the Sea Harrier, he fired a long burst at Ward. Piuma stopped firing when he too realised he did not have tracers. The Sea Harrier entered a valley and Piuma thought Ward would not be able to get away from him. At that moment, he was flying at 450kt and was now 120ft (37m) above the ground. Piuma saw a pilot hanging from his parachute (Senn). An instant later, he felt a powerful explosion rock his aircraft and he ejected. Ward and Thomas had seen the Dagger flying lower than them and heading northeast. Ward had got under and behind the tail of the last Dagger and fired his missile. The aircraft cartwheeled after exploding. Piuma had overtaken the Sea Harriers and had been fired upon by his intended target.

After resuming their patrol, the two Sea Harrier pilots lost sight of each other. Ward was afraid that Thomas had been shot down, as he was not answering his calls. Upon returning to HMS *Invincible*, Ward discovered that Thomas had just landed, his aircraft damaged by three rounds that had hit the electronics bay in the tail area. Apparently, Thomas had been hit by Argentine Army machine-gun fire while flying low over Fox Bay.

The downed Argentine pilots spent the night in the open. Piuma was injured and barely able to walk, Donadille's eyesight had been affected by the ejection and Senn was in good physical shape. They would spend between one and two days in the open before Argentine forces rescued them.

These were the squadron's first losses and were very keenly felt. Not only was the fate of the pilots unknown, but also the fleet had been reduced to C-401, C-420 and C-434. C-412, C-415 and C-432 had been taken out of service – the last with engine problems, while the other two were sent to Tandil to change their windscreens on 24 May and would return four days later. There was also a lack of 1,300-litre tanks, with only three left at San Julián. In light of this, C-419 was deployed to San Julián on 22 May. In fact, 1st Lieutenant Musso had intended to fly C-419 to Río Grande, but the pilot received the order to remain at San Julián to reinforce the squadron. The following day C-410 and C-416 arrived at San Julián. Also on 23 May an Aerolíneas Argentinas Boeing 707 arrived with a batch of 1,300-litre auxiliary fuel tanks, purchased from the Israeli government and delivered via Peru. Included in the shipment were new fuses, of the Kappa E type, which could be programmed to detonate after 2.6 seconds or 3 seconds. Thanks to these new fuses, more bombs exploded when hitting their target. The reinforcements were completed with the arrival of C-411 with 1st Lieutenant Gabari Zoco and C-430 with Captain Demierre from Tandil, and C-421 from Río Grande. However, C-411 went off the runway on landing and remained out service.

Bad weather impeded missions on 22 May, but the following day, the weather improved after midday, and Coral flight was ordered to take off by OF.1216. Coral flight consisted of Captain Dimeglio (C-421), 1st Lieutenant Román (C-434) and Lieutenant Aguirre Faget (C-420). At 12.30 the Daggers left, armed with two BRPs each and tasked with attacking ground targets near San Carlos. The aircraft delivered their bombs and returned without problems, but without knowing the effect of their attack.

The yellow identification stripes had not had the intended effect, and only increased the chances of detection by the Sea Harriers; their removal was ordered on 24 May. In the small town of San Julián, technicians only found green paint that, when applied over the yellow stripes, left a turquoise colour. On the same day, missions were ordered over San Carlos and Ajax Bay, in order to attack support ships. Escuadrón II received OF.1227 and OF.1228, which led to the formation of Plata and Oro flights. The first of these took off at 10.20, with Captain Dellepiane (C-434) and 1st Lieutenant Musso (C-420) and Callejo (C-421), to bomb ground targets with two BRPs each. Four minutes later Oro flight took off with Major Puga (C-410), Captain Díaz (C-419) and Lieutenant Castillo (C-430), this being tasked with attacking ships. According to 1st Lieutenant Callejo, just after he dropped his bombs (one in fact remained on the aircraft) over orange fuel tanks on the north shore of San Carlos Bay, he turned left to look for Cape Dolphin. After finding it, he continued in a turn to the left, heading west. Just as he was about to complete his crossing of Falkland Sound/San Carlos Strait, Callejo saw a frigate to his right, just north east of Pebble Island (this was the SAM trap consisting of the frigate HMS *Broadsword* and destroyer HMS *Coventry*). A missile was fired against him and Callejo faced the threat, turning right towards the missile and jettisoning his fuel tanks. After avoiding the missile, he changed his turn to head inland, flying supersonic. He made his way west, over-flying Borbón/Pebble Island and the rest of the islands north of Gran Malvina/West Falkland. Once he had passed these islands, he slowed and climbed in order to save fuel for the flight home. The other two aircraft of Plata flight made their return, flying over Gran Malvina.

While this was happening, Oro flight was approaching the zone to attack the southern part of the bay. Captain Raúl Díaz remembers that *'the first flight planned their attack beginning with a very low flight over Gran Malvina/West Falkland. My flight didn't follow them because we would have had to make a very tight, 150-degree turn inside the bay to reach the attack course, and we would have been an easy target for the anti-aircraft guns on the ships and the weapons on the coast. We decided to approach from the north of the island, reaching San Carlos Strait/Falkland Sound and approaching the target on a direct course at very low altitude. We would have to pass Sebaldes/Jason Islands at less than 50ft (15m), at a speed of about 480 to 520kt, in a lateral formation, to make an attack with the three aircraft simultaneously.*

'Our plan was accomplished, and when we passed Jason Islands we heard on our radios about the attack at San Carlos by the first flight. Our heading was about 100 degrees. Lieutenant Castillo was to my left, at about 2300m (7,546ft) and Major Puga to my right, at the same distance. When we were close to Elephant Bay we readied our weapons panels and prepared to start our final turn towards the target in 80 seconds. Focused on our instruments and watching the north end of the sound to see if there were any British ships, we didn't realise that two Sea Harriers were intercepting us from behind, guided by a frigate we couldn't see.

'The leader of the British pair launched a missile against Lieutenant Castillo and his aircraft exploded. Major Puga shouted 'Number 3 was shot down by a missile!' I thought he was talking about one of the aircraft of the other flight, which was over the landing zone. I continued my navigation. Immediately Puga shouted again 'Oro 3 was hit by a missile!' I reacted by looking to my right, where Puga was – in fact he was number 3 and Castillo was number 2. I saw he was intact but 200m (656ft) behind him an intense light was approaching at very high speed on a zigzag path. I realised it was a missile and that he had no time to react, so I shouted at him to eject. The missile hit his engine and the explosion was so spectacular that the smoke and fire completely covered the aircraft from about one metre behind the cockpit. I ejected the extra tanks and the bombs and started a tight turn to my right to see what was happening to Puga, who I was still telling to eject. When I was in the middle of my turn I felt a powerful shaking of my aircraft. I lost my controls and immediately all the warning lights on the panel went on. Diaz passed over Puga's aircraft and his aircraft straightened, so he ejected. He fell over Borbón/Pebble Island and Puga fell over the sea, but he made it to the coast and both were rescued. Castillo couldn't eject and went down with his aircraft'. The Daggers were shot down by the CAP formed by Lieutenant Commander Andy Auld (Sea Harrier XZ457) and Lieutenant David Smith (Sea Harrier ZA193).

David Smith remembers the combat. *'Andy Auld and I were on CAP to the north of the Falkland Sound under the control of HMS Coventry and HMS Broadsword. We had literally just arrived on task from HMS Hermes so were quite well off for fuel. HMS Coventry and HMS Broadsword were just north of Cape Dolphin. A few minutes after arriving on station the controller announced that the force had gone Air Raid Warning Red and to stand by. Several seconds later orders came to vector west at full speed to intercept inbound targets, strength unknown but believed to be three or four.*

'We increased speed to about 540kt and dropped down to 200ft (61m) over the sea, heading approximately 270 degrees. The controller was giving us 'Bravo' control,

which is essentially range and bearing of the target, and it was up to us to sort the geometry out and self-position for the intercept. At about 50 miles (80km) contact was lost by one or other of the ships but we kept heading and speed on the off chance of making an autonomous pick up. As it happened, contact was regained at about 40 miles (64km) and the B control continued. Andy led the left turn for the intercept and called 'Tally' at about five miles. I was in a fighting wing on his starboard echelon. He fired first at the left-hand Mirage (sic) flown by Lieutenant Castillo. He then quickly shifted aim and release his second missile at Oro 3 flown by Major Puga. The leader, Captain Diaz, then went into a hard right turn, clearing his wing in the process, and was hit by my missile as he was turning through about south.

'I was visually tracking him as his burning aircraft descended towards the high ground south of Pebble Island and wondering why he hadn't ejected. Moments before the impact several things happened. I saw the ejection sequence begin with the pilot pulling an orange parachute out of his cockpit (when the pilots ejects, a small parachute opens immediately after, this serving to decrease the speed of the seat before deploying the pilot's parachute).

'I heard the emergency beacon on 243 MHz. At the same moment a fourth Mirage (the Dagger of 1st Lieutenant Callejo of Plata flight) flashed under my nose heading west at high speed and I called a break towards it to try and get another missile off. As a result of this I only saw the fireball on impact of Diaz' Mirage out of the corner of my eye and never saw his fully deployed parachute. The fourth Mirage quickly out-distanced me, and it wasn't even worth trying a missile shot at the rapidly opening range.'

Once again, the squadron lost a complete flight and their commanding pilot, complicating the situation even more. C-434 was put out of service after returning with a broken windshield. However, C-432 was returned to service.

No operational sorties were made on 25 May and only C-401 flew, returning to Tandil for repairs. This left only four operational aircraft at San Julián. On the following day Poker flight was sent under OF.1241 to attack ground targets at San Carlos. The aircraft were carrying four BRPs and two 1,300-litre tanks. The flight departed San Julián at 13.30, and Gustavo Aguirre Faget remembers the mission. *'The only mission for which we volunteered was a dive-bombing sortie we made with Captain Dimeglio over San Carlos. We told our boss that because of the damage we had caused them, the British would be forced to spread their AAA defences. We told him that maybe it wasn't too dangerous to attack from a great height, and that we could test this by diving from 40,000 or 35,000ft (12,192 or 10,668m) to 12,000 or 15,000ft (3,658 or 4,572m), release the bombs, and then escape. The squadron commander said, 'OK, you will go tomorrow.*

'We flew on the following day, 26 May, when the beachhead was well established and there were a lot of ships. We supposed that at that height, with good visibility, we would have time to see the missiles from very far. Gran Malvina/West Falkland had perfect visibility, but from San Carlos Strait/Falkland Sound it was all completely covered by clouds. Anyway, we began our descent, diving at 60 degrees, almost without power, airbrakes open, one next to the other. When we were approaching 16,000 or 18,000ft (4,877 or 5,486m) Dimeglio said 'Let's go'. He stopped diving and began to return to San Julian but I kept diving. On that mission we only carried two 1,300-litre tanks and four 500lb (227kg) bombs with SSQ fuses, which exploded

on the surface, before penetrating, and offered improved aerodynamics. As we flew high, we had enough fuel. I followed and dropped the bombs without seeing the target. We established communications with the ground controller, who confirmed we were in the place for the attack. The target was close to San Carlos settlement. I don't know where the bombs fell.'

Dimeglio ejected his bombs on the return, close to Jason Islands, and at 15.20 the Daggers landed. Aguirre Faget recalls that *'during the night, the chief of Grupo 6 de Caza called. He was at Comodoro Rivadavia, with the Fuerza Aérea Sur, and I thought he would punish me for dropping the bombs. Instead he punished Dimeglio for not dropping the bombs over the target, because we were over it, and he ultimately dropped them over the water.'*

Final missions

The bad weather and the few serviceable aircraft prevented any new combat missions until 29 May. Meanwhile, aircraft had been repaired and eight Daggers were serviceable that day. Because of the fighting at Goose Green, on the morning of 29 May OF.1264 was ordered, calling for an attack on ground targets in the Darwin and Goose Green areas. Patria flight was prepared, with Captain Dimeglio in C-420, 1st Lieutenant Román in C-421 and Lieutenant Aguirre Faget in C-416. The third aircraft could not take off due to a failure of the oxygen system. The other two took off at 10.00, but the Comando de la Fuerza Aérea Sur (CdoFAS, Southern Air Force Command) could not coordinate the attack with the ground forces, which had surrendered during the night. The aircraft were ordered to return and landed at 11.00.

At 13.20, Puma flight (OF.1269) took off, comprising Captain Dellepiane (C-421), 1st Lieutenant Callejo (C-420) and Captain Luis Demierre (C-416). These flew together with León flight (OF.1270) with 1st Lieutenant Román (C-432) and Lieutenant Aguirre Faget (C-412). Just after the first aircraft passed over Borbón/Pebble Island, Dellepiane saw a Sea Harrier CAP approaching to intercept them. He ordered the Daggers to eject their external stores and return. When they turned back they saw hits on the water and supposed these were the Sea Harriers firing their guns. The Daggers selected afterburners and flew at Mach 1.2 until they reached Sebaldes/Jason Islands. When they ejected their tanks, Dellepiane's jet only jettisoned a single tank. Aguirre Faget remembers that *'when Román heard on the VHF radio the shouts of the pilots of Puma flight he asked their leader what was happening. Number 3, Captain Demierre, answered 'Return!' so we turned back. On the return we almost had an accident. When we were flying over Sebaldes/Jason Islands, at very low altitude, Román dropped his bombs instead of simply ejecting them, so as to test the fuses. He asked me if I was at a good distance and I said yes, but I forgot I was carrying the Kappa E fuses that detonate after three seconds. I remembered this when the parachutes of the BRPs were opening, so I descended further and saw the explosion at only 500m (1,640ft) to my side – it was a great stroke of luck that the debris did not hit us'*. All aircraft returned safely to base.

On the following day, Vicecomodoro Luis Villar was named as commander of the squadron, succeeding Sapolski, and on the following day a test was made with the Mk 17 bombs on the ventral pylon of C-432, with Captain Dimeglio at the controls.

Horacio Mir González, César Román and Gustavo Aguirre Faget, three Falklands War veterans, together with C-412, which participated in the first air strikes on 1 May. (Author's archive)

Bad weather again prevented missions and the squadron remained grounded until 4 June, when OF.1277 arrived and Piña flight was prepared. This consisted of Vicecomodoro Villar (C-432), Captain Demierre (C-420) and 1st Lieutenants Román (C-416) and Musso (C-421). The aircraft took off from San Julián at 15.00 carrying four BRPs each to attack ground targets. The mission was flown together with the Canberras of Lince and Puma flights. The Daggers flew at 36,000ft (10,973m) and arrived at a predetermined position. The ground radar operator then instructed them to turn 320 degrees. When they were 6 miles (10km) from the target they began their dive-bombing runs, one after the other, separated by about 1000m (3,281ft). At 20,000ft (6,096m) they dropped their bombs, but numbers 3 and 4 released only three weapons. The pilots could not verify the effects of the attack but returned safely to base.

On 5 June, Nene flight (OF.1283) took off for an armed reconnaissance over 9 de Julio/King George Bay, armed with guns and two BRPs each. At 14.40 Captains Maffeis (C-421) and Demierre (C-416) and 1st Lieutenant Musso (C-432) took off and were guided by the BAe 125 LV-ALW of Escuadrón Fénix until they were 140 miles (225km) from the zone that they were to reconnoitre. The flight could not find any target and, after dropping their bombs, they returned and landed at 17.00. Between 5-9 June, Captain Dimeglio and 1st Lieutenant Román went to Tandil to begin preparing the Mirage 5Ps received from Peru, together with AS.30 air-to-surface missiles. Peru offered to send the aircraft directly to San Julián and to use their own pilots on combat missions, but the offer was rejected due to the possible political consequences.

When the attack on Bluff Cove took place on 8 June, Escuadrón II received the order to fulfil diversionary missions to frustrate the CAPs. Carta flight (OF.1293) took off at 13.20, followed by Sobre flight (OF.1294). The first comprised Vicecomodoro Villar (C-411), Lieutenant Daniel Valente (C-412) and 1st Lieutenant Callejo (C-432), while the second included Captain Maffeis (C-416), 1st Lieutenant Musso (C-420) and Lieutenant Aguirre Faget (C-421). All aircraft were armed with guns only. After flying over the Sebaldes/Jason Islands, simulating an attack formation, they returned to BAM Gallegos, where Escuadrón II was deployed that day. Meanwhile, the remaining aircraft were flown there supported by C-130E TC-61, which carried technical personnel. The transfer was completed the following day.

Six Daggers were prepared with two BRPs each early on 10 June, before the order was changed and only two aircraft remained on alert with two Shafrir II missiles and 500-litre tanks, but no missions were flown. On the following day Dimeglio conducted tests with four locally built 125kg (276lb) Exocor bombs, launching them successfully from C-432 over the Gallegos estuary. Three Peruvian Air Force officers arrived at the base on 12 June to instruct on the use of the Mirage 5P, in particular in regards to the different equipment carried. It was planned to start using the Peruvian aircraft on combat missions at short notice, but the end of the war frustrated these plans.

The final missions by 'La Marinette' took place on 13 June, when OF.1317 and 1318 arrived. The first flight consisted of Captain Maffeis (C-411), 1st Lieutenant Callejo (C-420) and Lieutenant Valente (C-416). Their jets took off at 11.00 but Callejo was forced to return due to undercarriage problems. They were followed 10 minutes later by Gaucho flight, with Captain Dimeglio (C-432), 1st Lieutenant Román (C-421) and Lieutenant Aguirre Faget (C-412). Faget had to abort the mission due to a brake failure. The aircraft arrived from the south, along the coast of the islands, and continued to the northeast, with the islands on their left. When the first flight was approaching the target – ground positions close to Puerto Argentino/Stanley – they saw a helicopter and then a CAP flying very high. The leader ejected his bombs and the wingman ejected his tanks, and both aircraft returned to base. When the Gauchos were approaching, they found Royal Navy Lynx XZ233 from HMS *Cardiff*, to the south of San Carlos Strait/Falkland Sound. They attacked, but the helicopter made evasive manoeuvres and escaped. Each aircraft fired their guns during two passes, but the Lynx avoided them and returned to its ship. The Daggers returned to base at 13.00 and did not fly again during the conflict.

In the course of the war, the Daggers of Escuadrón II dropped 84 250kg (551lb) Expal bombs and expended 2,920 rounds of 30mm ammunition. On 25 June they returned to VI Brigada Aérea, having lost six aircraft and one pilot.

Escuadrón III Aeromóvil

The first eight aircraft (C-410, C-421, C-428, C-429, C-430, C-431, C-433 and C-436) for Escuadrón III Aeromóvil arrived at Río Grande on 7 April. Here, the squadron was established under the command of Major Carlos Martínez. On 17 April, Captain Mir González (C-429) and Lieutenant Bean (C-430) conducted an exercise together with other FAA aircraft, making a simulated attack against ships of the Argentine Navy, guided by Lockheed SP-2H Neptune 2-P-112. The two Daggers departed Río Grande at 11.00, and after looking for their targets without success, they returned to base at 13.00.

On the same day C-421 was sent to Tandil for repairs, followed the next day by C-428. C-435 and C-437 replaced these aircraft.

Since they shared a base with the Super Etendards of the Argentine Navy, on 25 April a joint exercise was prepared, with two Super Etendards armed with Exocet anti-ship missiles and two Dagger escorts. On the take-off one Dagger experienced technical problems, while the other suffered a fuel system failure just after taking off. The mission was cancelled. Another mission was made that day, however, this consisting of a flight over the islands by Major Martínez (C-414) and Captains Rhode (C-435), Cimatti (C-429) and Robles (C-436).

The first operational sortie was planned on 30 April, when two Daggers manned by Captain Moreno and Lieutenant Volponi were tasked to cover an Argentine Navy

SP-2H Neptune on a maritime surveillance flight close to the islands, to be executed during 1 May. The British attack on BAM Malvinas changed these plans.

For the first day of the war it was decided that the Daggers at San Julián would mainly fly attack missions and those from Río Grande would fly air cover, each armed with two Rafael Shafrir II air-to-air missiles, guns and 1,300-litre tanks. At 07.30 on 1 May OF.1091 arrived, ordering the departure of Toro flight with Captain Moreno (C-437) and Lieutenant Volponi (C-430). Captain Carlos Moreno remembers that mission.

'At 07.45 we were on the end of the runway with Héctor Volponi. We released the brakes for our first combat mission. The weather was very bad, with rain and a very low cloud base.

'At about 08.25 we were in contact with the Malvinas radar, and were 50 miles (80km) from Puerto Argentino/Stanley. The radar operator told us that the airport was under attack and that a Sea Harrier was falling in flames over the harbour. At the same moment he informed us that he had two 'pigeons' for us. I asked what their position was and he told me they were 138 miles (222km) from us. '020 degrees', I answered and ordered Volponi to select full power. Volponi was to my left, about 500m (1,640ft) to my side and 10 degrees behind me.

'Heading 030 degrees', ordered the radar operator, 'the enemies are at 30 miles (48km) on reciprocal'. I asked what reciprocal meant, as I had either never heard that word or I didn't remember what it meant. He answered 'Head on, boy, they are coming from the front!' Now I understood and asked him to put me to one side of them, because I knew they could fire their AIM-9L Sidewinders from the front and we couldn't.

'No, you have them at 9 miles (14km), from the front and one mile to the right side!' I ordered the wing tanks to be ejected, and we only kept the ventral tank. I asked the operator to ask us for our fuel readings during the combat, because I didn't think we would remember to check in the heat of the moment.

'When the radar operator said we were crossing we ejected the last tank, selected afterburners and turned to the right. The tanks passed very close to the Sea Harriers and their pilots thought they were missiles. We had 22,000ft (6,706m) and they were at 18,000ft (5,486m), the radar operator told us. We started to fly in circles, trying to see them, and I asked the radar operator to tell us where they were. He answered 'In a circle, four aircraft, and I cannot recognise them!

'Desperately we searched for the Sea Harriers but we couldn't see them. Then Héctor asked me if I had fired a missile. I answered no and he said he saw a missile passing between our aircraft.

'Meanwhile the radar operator asked us about our fuel and at one moment he said we were behind the Sea Harriers, but then they were behind us, as they had turned very fast. I don't know how much time had passed, but I think it wasn't more than two minutes, and then we had 2,450 litres (539 Imp gal, the minimum required to return to base). I ordered Héctor to climb with afterburners in a very tight turn and we headed for home. The operator told us the enemy were also leaving. We climbed, Volponi to 37,000ft (11,278m) and myself to 36,000ft (10,973m), as we couldn't see each other. The radar operator gave us our position relative to Puerto Argentino – we were 5 miles (8km) from there. We started to calculate the return, as we were with minimum fuel and far from home, but we had only a mild wind against us and we could reach the base, although it was almost at the limit. We landed safely.'

The two Daggers had encountered Sea Harrier ZA175 flown by Lieutenant Commander Robin Kent and Sea Harrier XZ498 of Lieutenant Brian Haigh. The two aircraft were from 801 Naval Air Squadron, based on HMS *Invincible*. The Daggers landed at 09.45 after having flown the first FAA combat mission of the war.

The first flight of Daggers was followed at 10.00 by Limón flight (OF.1099), with Major Martínez (C-435) and 1st Lieutenant Luna (C-429). These were first vectored to a target that turned out to be the A-4Bs of Topo flight. Limón flight was then guided to a CAP of Sea Harriers, which they crossed at different altitudes. After manoeuvring for some seconds, both flights gave up the fight without using their weapons. The Daggers landed at 12.15.

Another two flights took off almost immediately: the first at 12.30 was Ciclón flight (OF.1100) consisting of Captain Mir González (C-430) and Lieutenant Bernhardt (C-437). They were guided against a Sea Harrier CAP and entered combat. When Bernhardt realised the Sea Harriers were reaching an advantageous position and descending, he departed in a climb and the British broke off the combat.

Rubio (OF.1113) followed this flight at 15.54, with Captain Carlos Rohde and Lieutenant Ardiles (C-433). The leader could not depart because of technical problems, and Ardiles flew alone. This was a big mistake, as he was an easy target for the Sea Harriers. Reaching the islands, the radar operator guided him to a radar echo, but this immediately become two echoes. These were the Sea Harriers of Lieutenant Martin Hale and Flight Lieutenant Bertie Penfold. When they crossed from the front, Ardiles ejected his tanks, which were again erroneously identified by Hale as missiles, and the Sea Harrier pilot conducted evasive manoeuvres. Meanwhile, Penfold, who was flying far from his leader and was not being pursued by Ardiles, put his aircraft on the tail of Ardiles' Dagger and fired a Sidewinder at 16.41. The missile destroyed the Dagger and Ardiles could not eject. He became the first pilot lost by the unit.

The departure of three flights was planned for the following day, using the seven aircraft available (C-410, C-429, C-430, C-431, C-435, C-436 and C-437). These were intended to attack ships, but because the British stayed so far from the islands, no missions were launched. Meanwhile, C-414 was sent from Tandil to make good the loss of C-433.

On 3 May, Daggers took off again on CAP missions, after receiving OF.1154. Dardo flight included Captain Mir González (C-437) and 1st Lieutenant Luna (C-435). They took off at 15.30 to protect planned A-4 attack missions. However, the enemy was not detected, and the Daggers returned two hours later.

When the Carrier Battle Group was detected on 4 May, an attack mission was planned with the Super Etendard, which was to refuel from a KC-130H. To provide air cover, OF.1161 was issued and Pollo flight was prepared. This comprised Captains Amílcar Cimatti (C-437) and Higinio Robles (C-414). The Daggers took off at 10.20 armed with Shafrir II missiles. Following the Super Etendard mission, they returned at 13.00.

In the afternoon, SP-2H Neptune 0707/2-P-111 attempted to verify the damage caused by the Super Etendard mission. At 15.52 a message from a British ship was intercepted, revealing that they had detected the Neptune and were requesting a CAP to be sent to shoot down the Argentine aircraft. Immediately, Talo flight (OF.1163) was dispatched, taking off from Río Grande at 16.00. After protecting the return of the Neptune, the Daggers landed at 17.00 without having found the Sea Harriers.

Due to the lack of 1,300-litre tanks, the use of French-made 1,700-litre tanks was tested on a Mirage IIIEA on 5 May.

On 6 May it was planned to send Cobra and Pitón flights (OF.1181 and 1182), with five aircraft each armed with a single 1,000lb Mk 17 bomb. This was part of a massive attack planned against the British fleet, if they were to approach the islands. The attack force also included Skyhawks. However, the British were not found, and the aircraft remained on the ground. Aguila flight (OF.1173) had taken off at 11.40 to cover the planned attack. Aircraft C-430 and C-437, flown by Captains Cimatti and Robles, were armed with Shafrir II missiles. When they began their return they test-fired the guns. On C-437, the system that was supposed to reduce engine power when the guns were being fired remained engaged, and the power reduction meant the aircraft began to lose height. Robles ejected his ventral tank and descended from 27,000 to 20,000ft (8,227 to 6,096m). He then received the order to return to the islands and eject. Some minutes later the engine worked again properly, so he decided to continue to the mainland where he landed at 13.20.

A formation of five ships was detected sailing near Isla de los Estados, close to Tierra del Fuego and Río Grande, on 7 May. In response, three Daggers were sent from Río Grande to Comodoro Rivadavia. A strike was planned with Super Etendards, but after the ships were identified as Soviet trawlers, the mission was cancelled.

No more flights were made until 12 May. On this day, departure of a flight of four Daggers, each armed with a single Mk 17 bomb was ordered. The aircraft were to attack the destroyer HMS *Glasgow* and the frigate HMS *Brilliant*, which were bombarding Argentine positions. In the event, three of the four aircraft had problems starting their engines, and the mission was cancelled.

C-418 was deployed from Tandil on the following day, and C-428 also returned to its deployment base. Meanwhile, Captain Rohde and Lieutenant Volponi made a patrol flight over the islands between 16.30 and 17.30, but located no targets.

Aircraft C-414 and C-430 returned to Tandil on 14 May and test flights were made with C-418, C-428, C-435 and C-437. Another mission took place the following day. This flight consisted of Captain Rohde (C-418), Captain Janett (C-428) and Lieutenant Bean (C-435), each with a single Mk 17. The aircraft took off at 16.00 and landed at 17.30 without having found any targets.

Bad weather conditions and the difficulties in finding the British fleet, which was sailing to the east of the islands, led to the grounding of the squadron during the following days. Meanwhile, on 18 May, C-401 arrived from San Julián.

On 19 May Major Martínez (C-431) and Lieutenants Bernhardt (C-418) and Volponi (C-435) flew another attack sortie between from 15.45 and 17.45, but they failed to find any targets. C-409 was also deployed in this period, providing a total of 10 aircraft at Río Grande.

Bomb alley

The expected British landing took place on 21 May and numerous attack missions were launched against it. Among the first was Ñandú flight (OF.1181) and Perro flight (OF.1182), which took off from Río Grande, providing Daggers each armed with a single Mk 17 bomb. The first flight included Captains Rohde (C-409) and Janett (C-436) and Lieutenant Bean (C-428) and took off at 09.45. They were the first aircraft to arrive at San Carlos from mainland bases. When they arrived in the sound at 10.25, Rohde left his wingmen and attacked the frigate HMS *Argonaut*, without hitting her. The other two went for HMS *Broadsword*, but when they were on the final run, the frigate fired a Sea Wolf missile at Bean's aircraft, which disintegrated over the water. The pilot ejected but was not rescued. John Keith Watson, chaplain of HMS *Broadsword* remembered the attack. *'I was on the bridge and saw a dot coming over the hills, which expanded into two dots, then our Sea Wolf system fired, taking out one of the dots, which by now was an aircraft, and the pilot parachuted out. The second aircraft fired at us with cannon, and then was away over the hills. One minute later, I saw another aircraft, but nothing more, as my head was on the floor. People had been injured at the rear of the ship'*. The other two Daggers returned home and landed safely at 11.45.

Perro flight included Captain Moreno (C-437), Lieutenant Volponi (C-418) and Major Martínez (C-435) and took off immediately after the others. On arriving in the sound, they saw the destroyer HMS *Antrim*, recently attacked by León flight from San Julián. The Daggers fired their guns and hit the destroyer with a Mk 17, but the bomb failed to explode. The destroyer fired a Sea Slug missile without aiming, as a last attempt to confuse the attackers, but to no effect. The bomb passed through the Sea Slug magazine but failed to explode and was later disarmed. The *Antrim* was put out of combat for some days and was replaced by HMS *Glamorgan*. As the Daggers were leaving the sound, the frigate HMS *Brilliant* guided the CAP of Lieutenant Commander Rod Frederiksen and Lieutenant Hale to intercept them. The Sea Harrier pilots saw the Daggers and Hale fired a Sidewinder almost out of range. The missile failed. The Daggers made their escape thanks to their higher speed, and landed at 11.45.

Once the results of the morning missions were known, a second wave of attacks was prepared, in which the 'Avutardas Salvajes' took part with two flights. The first was Cueca flight (OF.1193) with Captain Mir González (C-418), Lieutenant Bernhardt (C-436) and 1st Lieutenant Luna (C409). The second was Libra flight (OF.1194) with Captains Cimatti and Robles (C-429). The aircraft began their take off at 13.55 but Cimatti suffered an oil leak after launching, leaving a dense black contrail. Cimatti returned to base while Aérospatiale SA.330 Puma PA-13 of the Prefectura Naval departed from Río Grande for a search and rescue mission. In the event, Cimatti made it home. Robles joined Cueca flight and continued with them. He remembers that *'We approached the sound from Salvajes/Jason Islands; we had to pass over the Hornby Mountains but they were completely covered by clouds and we couldn't pass.*

'We flew along the hills until we found a small valley through which we could see the other side. We went through it. Now we were only three as we supposed Luna crashed against a hill... On the other side of the sound, at Ruiz Puente Bay/ Grantham Sound we saw a frigate close to the coast. We started the attack, thinking 'For Luna!' We started flying very close to the water when they opened fire on us. Ahead of the aircraft of Mir González a path was formed by his bullets hitting the water. His risky flight took him through the frigate antennas. His bomb was 10m (33ft) short, raised a big mass of water that covered the ship, then bounced and hit the hull.

'Lieutenant Bernhardt dropped his bomb, which hit the upper part of the ship, towards the front. I came into firing range, dropped my bomb and when I was jumping the ship I saw a big part of the rectangular antenna passing close to my canopy

and turning in the air. To my right a missile passed, looking for Bernhardt's aircraft. I shouted to him to turn to the right, and the missile passed very close. While we were leaving, turning to the right, the frigate was only a smoke cloud'. The ship that had been attacked was HMS *Ardent* and Luna had not hit a hill, as the pilot himself remembers: *'at only three minutes to the target about five Sea Harriers intercepted us. Because of the bad weather we had to get into a hill one by one and I was the last. When I got in I saw a shadow passing above me and to my left. It was a Sea Harrier going for Captain Robles' aircraft. Almost simultaneously I saw fire in my mirror and immediately I felt the impact of a missile in my aircraft, which remained without controls. I tried to climb and the aircraft turned upside down, I thought I would die. I left the stick and searched desperately for the upper ejection handle. I think I found it when I was again flying normally, as the ejection was normal.'*

The Daggers were detected over King George Bay by HMS *Brilliant*, which sent the CAP of Lieutenant Commander Rod Frederiksen (Sea Harrier XZ455) and Lieutenant Andy Auld (Sea Harrier ZA176). The Sea Harriers saw the Daggers and Frederiksen fired the Sidewinder that hit Luna, who was rescued by islanders. The *Ardent* received the impact of the Mk 17 bomb dropped by Bernhardt, which completely destroyed the hangar and ship's helicopter, as well as the Sea Cat missile launcher. Robles's bomb hit the ship but did not explode.

For the squadron, the day was coming to an end, with two aircraft and one pilot lost. As a result of the losses, on C-414 and C-417 were deployed the following day, but C-410 was sent to San Julián. No sorties were made because of bad weather.

Activities resumed on 23 May, when Puma flight (OF.1205) and Potro flight (OF.1206) were dispatched. The first comprised Captain Cimatti (C-417), 1st Lieutenant Jorge Ratti (C-418) and Captain Rohde (C-436). The second included Captain Carlos Moreno (C-414), Lieutenant Héctor Volponi (C-437) and Captain Higinio Robles (C-435). They took off at 08.45 and were to be guided by Learjet T-23. However, they could not find the Learjet due to poor visibility and had to return.

The weather improved in the afternoon and 14.20 saw the departure of Daga (OF.1214) and Puñal (OF.1215) flights, each aircraft carrying a single Mk 17 bomb. Captains Cimatti (C-417) and Rohde (C-414) and 1st Lieutenant Ratti (C-418) formed the first flight. The second comprised Major Martínez (C-429) and Lieutenant Volponi (C-437), while Captain Moreno could not take off because of technical problems with his engine. Just after take-off, Cimatti had to return because of a fuel leak. When the Daggers arrived at San Carlos they could not find any target, and almost collided with a flight of Daggers from San Julián. Daga flight then returned to base.

Meanwhile, Puñal flight flew over San Carlos without finding targets. As they were returning they were intercepted by the CAP formed by Lieutenant Commander Andy Auld and Lieutenant Martin Hale (the latter flying Sea Harrier ZA194). The Daggers saw them, jettisoned their bombs and engaged afterburners. Hale saw Martínez and began the pursuit, but the Dagger was almost out of range and, thanks to its greater speed, was increasing the distance. Hale feared a repeat of the situation he had experienced two days earlier, when he had found the same pilots, but the Sidewinder fired against Volponi missed the target since it was out of range. As he watched Martínez depart, the Dagger of Volponi came into view, about a mile closer than his leader. Hale managed to engage Volponi's aircraft and fired a missile before the he could put himself out of range. The Sidewinder hit its target and Volponi could not eject.

On 24 May the missions over San Carlos were repeated, this time targeting the transport vessels at Ajax Bay. The first mission from Río Grande was flown by Azul flight (OF.1225), each Dagger armed with a single Mk 17 bomb. The flight was led by Captain Mir González (C-436), with Captains Robles (C-418) and Maffeis (C-431) and Lieutenant Bernhardt (C-417) as wingmen. The aircraft crossed the sound and although they reported that a Sea Harrier fired on them with guns, there is no British report to verify this. Once over Bahía de Ruiz Puente/Grantham Sound, the Daggers turned to the north and entered Ajax Bay from the south. C-431 did not release its bomb and so Maffeis attacked the RFA *Sir Bedivere* with guns. Here he describes the attack. *'Contrary to my three wingmen, I went over the eastern shore where there was a major accumulation of ships. On the west side there was a big transport. My decision to attack smaller vessels was based on the fact that, four minutes before, the generator on C-431 had failed, and with this breakdown the bomb could not be dropped. It was a failure of the aircraft's electrical circuit and I could only fire the two 30mm guns. I pulled the trigger three times: the first shots were short and raised big plumes of water, but the next may have impacted on the target.*

'All this happened very quickly but, as in other engagements, it remains firmly etched in my mind. When the gun camera films were developed they revealed – in addition to such technical details as a 350-degree attack heading – six ships, or about 50 per cent of all British ships in San Carlos Water.'

The other aircraft attacked a ground target and when they were approaching the entry to the bay, some ships opened fire on them. Able Seaman (M) Neil Wilkinson was responsible for the 40mm Bofors gun on HMS *Intrepid*. *'I do recall we were at the top end of San Carlos and the Mirages (sic) came around one after the other as if following each other. When I opened fire on them they were crossing down the starboard side of the ship but some distance away, my tracer was going in front or behind the aircraft, then I saw bits fly off either the tail or the wing, not sure which area it came from, then they were gone but the aircraft dipped as if the pilot was then struggling with the aircraft.'*

Information indicated that the British had installed a radar on Beauchene Island, a small and lonely outcrop located 59km (37 miles) to the south of East Falkland/Isla Soledad. It was therefore decided to send an air strike with Daggers. The raid was launched on 25 May after the receipt of OF.1233 at Río Grande. Rango flight was formed, with Captains Rohde (C-418) and Janett (C-431), and because the Daggers lacked navigation equipment capable of detecting the tiny island it was decided that they would be guided by Learjet T-23 until they were close to the target. The Daggers took off at 10.00 and located the island, but after flying over it many times, they found nothing and returned to Río Grande at 12.00. Meanwhile, OF.1234 took off at the same time, this being Bingo flight, with Captains Cimatti (C-436) and Moreno (C-435). Their task was to detect possible British positions on Belgrano/Meredith Cape, on the southern extremity of Gran Malvina/West Falkland. These Daggers were also to be guided by Learjet, but due to a problem with this latter aircraft, they went alone. When they saw the islands Cimatti ordered the Daggers to accelerate to 500kt and fire their guns over the rocks, to see if there was any AAA that might fire upon them. They saw nothing and after patrolling the southern part of the island, they returned to base and landed at 12.20.

The six aircraft at Río Grande did not make another mission that day. The following day C-429 was sent back to Tandil, while test flights were made with C-431 and C-417.

No more attack missions were launched until 28 May, when Poker flight was sent according to OF.1256, with Captains Maffeis (C-414), Ratti (C-436) and Janett (C-418) and Major Martínez (C-435). Poker flight was guided by Learjet T-23, and was tasked with attacking naval targets at Bahía de Ruíz Puente/Grantham Sound. The Daggers found nothing and returned to base.

Ñandú flight (OF.1266) was sent the following day, when Captain Mir González (C-414) and Lieutenant Bernhardt (C-436) took off at 11.30 for an armed reconnaissance over San Carlos. They were guided to the target by Learjet LV-LOG from Escuadrón Fénix. Although they did not see any ship or British defences, as they were leaving San Carlos a Rapier surface-to-air missile hit Bernhardt's aircraft and destroyed it, killing the pilot.

After this loss the unit was left with only four aircraft, but C-415 and C-401 (flown to Tandil for repairs) were returned by 1st Lieutenant José Gabari Zoco and Carlos Antonietti.

No more forward observers remained close to San Carlos and this, together with the bad weather, prevented any new mission until 5 June. At 13.30, Captain Ratti (C-415) and 1st Lieutenants Gabari Zoco (C-414) and Antonietti (C-417) took off for an armed reconnaissance to the west of the islands, but they found nothing and returned at 15.30. Meanwhile, Major Martínez (C-401) and Captain Moreno (C-418) took off on a similar mission at 14.00 took off. They also found nothing, and returned at 16.00.

At 15.15 on 7 June, Captain Rohde departed alone with C-401 for a test flight and returned one hour later. Meanwhile, at 14.45 Captain Mir González (C-418) took off with 1st Lieutenants Gabari Zoco (C-417) and Antonietti (C-415) for another armed reconnaissance. Finding nothing, they returned at 16.00.

Endgame

On 7 June the presence of British ships and ground forces at Fitzroy and Bluff Cove was detected and an air strike was ordered for first light on 8 June. After the departure of the A-4B Skyhawks at 11.30, two Dagger flights took off from Río Grande at 13.00. The first, with the Perro callsign (OF.1291), comprised Captain Rohde (C-415) and 1st Lieutenants Gabari Zoco (C-417) and Ratti (C-401). The second, with the Gato callsign, comprised Captain Cimatti (C-435), Major Martínez (C-418) and 1st Lieutenant Antonietti (C-431). After a bird hit his windscreen on take-off, Antonietti had to return. The five remaining Daggers continued, guided by Learjet T-23. Each aircraft was armed with two 250kg BRP bombs.

Lieutenant Jorge Ratti was number 3 in the first flight. *'We approached from the south and the approach to Bluff Cove was made over the coast. We passed a frigate entering the bay from the south. There was AAA. I saw a hit between number 1 and number 2 on the sea. Only one bomb was released from my aircraft due to an electrical failure. On the film, gun impacts could be seen on the ship.'*

José Gabari Zoco, the number 2, continues the story. *'The frigate was entering the bay. When we saw her to the south of our position, we turned almost 270 degrees to the left to face her from the land to the sea. When we made the turn, the frigate also turned 180 degrees and was trying to reach the open sea. I also launched my bombs but I don't think they hit the ship. The only aircraft to receive a hit was mine.'*

Gato flight attacked behind the Perros, and all its aircraft launched their bombs despite the attentions of the frigate, which fired 40mm Oerlikon guns, machine-guns and Sea Cat missiles. Although all the pilots' reports state they were to the south of Bluff Cove, the frigate attacked was HMS *Plymouth*, which was sailing on San Carlos Strait/Falkland Sound, close to Chancho Point. HMS *Plymouth* was serving as a radar picket to protect the support ships carrying supplies to the beachhead. The aircraft approached the islands from the south and turned to the north, flying through the sound. One bomb passed through the ship's funnel, without causing great damage. Another entered the hulk and did not explode, while a third damaged the anti-submarine mortar before falling into the sea. The last bomb hit the edge of the flight deck and caused a depth charge to explode before continuing its way to the sea. The HMS *Plymouth* also received 30mm gunfire from the Daggers. Five crewmembers were injured and the ship was put out of action. The destroyer HMS *Exeter* vectored a CAP of Sea Harriers that was over Fitzroy to try to intercept the Daggers. However, the Argentines made their escape before the Sea Harriers could close. Although the HMS *Plymouth* claimed to have shot down one or two Daggers, all aircraft landed safely at 15.00. The crew on the HMS *Plymouth* controlled the fires and the ship sailed to San Carlos Bay where it could be protected by land-based air defences and the other warships.

An armed reconnaissance mission was made on the following day, when a flight (OF.1304) took off at 14.45. Captain Mir González (C-417), Major Martínez (C-401) and 1st Lieutenant Antonietti (C-418) were to fly over Sea Lion Island, guided again by Learjet T-23. After finding nothing, they landed at Río Grande at 16.45. On 10 June Captains Ratti (C-401) and Robles (C-414) made another unsuccessful reconnaissance mission, between 14.00 and 16.00.

The last missions took place on 13 June, when Zeus flight (OF.1323) and Vulcano flight (OF.1324) were launched. The first flight comprised Captain Rohde (C-401), 1st Lieutenant Gabari Zoco (C-415) and Captain Moreno. The second included Captain Janett (C-418), 1st Lieutenant Antonietti (C-414) and Captain Robles. Initially the mission was planned as a low-level bombing raid over Port Harriet, to the southwest of Puerto Argentino/Stanley, but before take off it was changed to a high-level attack, so the fuses had to be changed.

As the fuses were being changed, the jacket of Suboficial Auxiliar Pedro Miranda became hooked on the fuse of one of the bombs on C-418, and the fuse started working. Miranda raised the alarm and managed to unscrew the fuse just before the firing was initiated, saving his life and the aircraft.

Robles and Moreno could not take off due to technical failures and only four Daggers departed Río Grande at 15.15. At 100km (62 miles) from the target Antonietti experienced problems with the engine and had to return. The other three Daggers joined up and approached the islands from the south, but when they were approaching the target, the ground radar operator informed them that a CAP was heading towards them. They aborted the mission and landed at Río Grande at 17.15.

Once the war was over, on 19 June aircraft C-401, C-414, C-415, C-417, C-418 and C-435 returned to Tandil, followed some days later by C-431. In the course of the fighting, the unit had lost five aircraft and four pilots. A total of 70 combat sorties had been launched, of which 42 reached their targets.

Return to Tandil, start of Finger programme

As soon as the war had begun, the British and Canadian personnel from Marconi and Ferranti who were working on the SINT project left the country. Part of the project was left in limbo, but the Argentine engineers kept working. In August 1982, the REI program (Reemplazo Equipamiento Inglés, British Equipment Replacement) began and contacts were signed with different companies to buy new equipment. The REI programme later received the name Finger.

In March 1983, test flights took place with the Finger I prototype (C-408). This aircraft included software developed by the FAA and a Thompson-CSF head-up display instead of the HUD from Marconi. Meanwhile the Finger II was under development, this having a French HUD and electronic unit, the latter replacing the Ferranti equipment, but this aircraft lacked the ADC. The Israeli radar was retained on all prototypes. Plans were made to modify the aircraft at ARMACUAR, but budget reductions took effect when Raúl Alfonsín assumed the presidency in December 1983. As a result, the programme was delayed. In the meantime, C-408 began firing trials in November.

Together with the 10 Mirage 5Ps that arrived from Peru, Argentina received 30 AS.30 air-to-surface missiles delivered, together with a pair of AS.20s intended for training. The Dagger's other weapons consisted of conventional bombs, Shafrir missiles and rockets, the latter including the LAU-51, LAU-61 and LAU-10 as well as the locally built Mamboretá rocket launchers. In practice, the Daggers used rockets only rarely.

In July 1983, while the conversion of C-405 to Finger II or 'semi-Finger' configuration was being completed, FAA's Comando de Material (Materiel Command) formulated a plan for the Dagger modification. This was to be completed according to different levels: Finger II, IIIA and IIIB. As 11 electronic units were available, 11 aircraft would receive this equipment, together with the Thompson-CSF HUD and the original software, and would become Finger IIIA aircraft. The Finger IIIB aircraft would receive the Thompson-CSF electronic unit and HUD, together with the fourth edition of the FAA-developed software and a new link between the radar and the Israeli ADC, in

C-408 was prototype for the Finger project, becoming first a Finger I and then a Finger IIIA. (Author's archive)

order that the latter could operate with the French equipment. The British Dopplers were retained, as there were enough to equip all the aircraft. According to the plan, the Finger IIIB was to be the definitive version.

In the same year, a simulator was received from Israel, this having been purchased as part of the first Dagger contract. On 19 April the Politronic towed target was tested, this being intended for use both on the Dagger and the Mirage IIIEA.

For 1984 the FAA had 26 Daggers, together with five Mirage 5Ps. The other five ex-Peruvian Air Force aircraft were in storage at ARMACUAR. VI Brigada Aérea also operated Aero Commander AC500U T-135 and FMA/Cessna 182J PG-341 and PG-348. The Daggers and Mirage 5Ps recorded 3,465 flying hours in 1984.

In the same year, the modified aircraft received a new Kfir-style nose and their intended onboard equipment, with the exception of the planned HUD. On 18 October C-427 made its first flight as prototype for the Finger IIIA. On 10 December the aircraft was delivered to VI Brigada Aérea together with C-422, similarly modified. Meanwhile, C-405 was at Area de Material Río IV to be upgraded from Finger II to Finger IIIA configuration, together with C-420. C-408 remained in Finger I configuration. As part of the homologation process, in 1983-84 the aircraft made a number of navigation and interception flights. By this time, the FAA had the equipment required to complete six aircraft to Finger I and II standard, and were waiting to receive the remaining equipment.

On 16 May 1985 another Dagger was lost, when C-431 crashed at 11.09 while pulling out of an air-to-surface attack on the Mar Chiquita firing range, in Buenos Aires province. Captain Norberto Cayetano Prior ejected safely.

The most prestigious event that year occurred on 4 September, when Major Mir González took president Raúl Alfonsín for a flight onboard C-438, taking off from Tandil at 10.00 using the Patria callsign.

The Daggers recorded 3,200.15 flying hours in 1985 and four aircraft received modifications for the Finger project. Plans existed for the installation of the remaining equipment in the near future, as this was already in the country. The four upgraded aircraft continued to make test flights as final adjustments were made to the configuration. In October, the FAA and Israel's Elisra defined the requirement for the development of a prototype Finger IV. This was to be a Finger IIIB with a radar warning

A Dagger with a new, white-painted nose. (Author's archive)

C-405 was a Finger II and later became a Finger IIIA. (Author's archive)

receiver, and C-412 was designated as the prototype. However, the modification was subsequently abandoned.

In 1985 the FAA conducted a second dissimilar air combat exercise, one of a total of seven to be held in the following years. Operativo Zonda 85 was a repeat of Zonda 84, and was held from 24-30 June. The exercise involved Grupo II de Vigilancia y Control del Espacio Aéreo (Airspace Vigilance and Control) and all brigades except I, II and X. The exercise included simulated air combat between different FAA types, including the Mirage IIIC/E, Mirage 5, A-4B/C Skyhawk, IA-58 Pucará, FMA/Morane-Saulnier MS.760 Paris and F-86F Sabre. The Mirage IIIC showed itself to be the most agile fighter at high speeds, and the most difficult prize for the Daggers was the Pucará. Thanks to the Pucará's low speed, the Dagger could not defeat it in a dogfight and instead had to make high-speed firing passes and then try to escape. Since the Pucarás always operated in pairs, there was always one providing cover for the other, and waiting to fire on the Dagger. On many occasions Dagger pilots flew in the rear seat of the Pucarás, and Pucará pilots did the same on two-seat Daggers and Mirage IIIs, in order to experience the capabilities of the opposing aircraft.

These exercises also included evaluation of the performance of the different Mirage versions. On one occasion, a Mirage 5 flew together a number of times with a Mirage IIIC and a Mirage IIIEA, and at 300kt they all selected full throttle. Each time, the Mirage IIIC was the first to reach 500kt. At 550kt the three aircraft began a 30-degree climb and the Mirage IIIC proved quickest in the climb.

The Finger operational

By 1986, Dagger C-423 had been updated to Finger IIIB standard, and Vicecomodoro Sapolski undertook air-to-surface training sorties to prove the system, dropping 100 125kg bombs. Captain Mario Callejo conducted air-to-air training with guns, followed in 1987 by tests with the Shafrir missile. In 1987 tests in which Zuni rockets were fired

Latin American Mirages

A Finger during a deployment to Patagonia in the late 1980s. (Author's archive)

from LAU-10 launchers failed, as the launchers experienced severe vibrations after the nose cone was released. These tests ended with the loss of the launchers, fortunately without causing major damage to the aircraft. After all these tests, the Finger was declared operational, and the FAA now had 14 complete modification kits.

On 13 November 1986 the Dagger celebrated its first 25,000 flying hours in FAA service, and the unit recorded 3,195 hours in the same year. On 7 November the Mirage 5Ps were transferred to X Brigada Aérea, leaving 25 Daggers/Fingers at Tandil.

The modification of all the aircraft to Finger IIIA and B standards began in 1987, and was completed that year, with 14 aircraft upgraded. The three two-seaters were not modified. On 12 June another accident took place, when C-418 crashed close to the base, Captain Fernando Robledo ejecting safely.

A Finger and a Mirage flying over the city of Tandil. The Finger is armed with six 250kg (551lb) Expal bombs and two Shafrir missile training rounds; the Mirage IIIEA carries two R.550 Magic and one R.530 missiles.
(VI Brigada Aérea)

Argentina

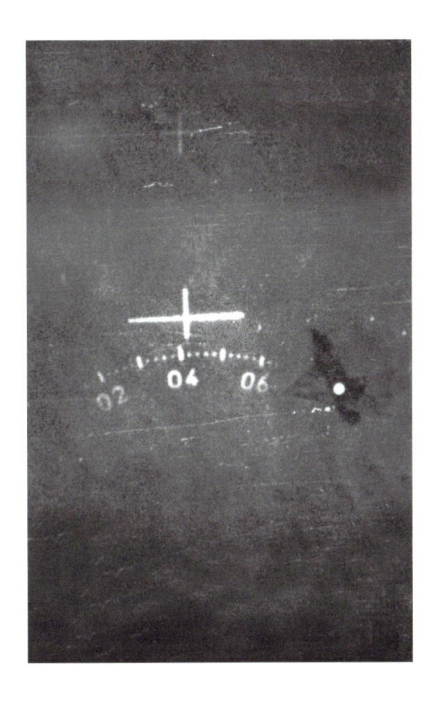

A Finger makes a snapshot on another Finger during air-to-air training.
(Comodoro Macaya)

As a result of the budget cuts and fleet reductions, in 1988 the Mirage IIIEA/DA jets of VIII Brigada Aérea were transferred to VI Brigada Aérea, forming Escuadrón II. The Mirage 5s, Daggers and Fingers meanwhile formed Escuadrón I. Later, on 19 November, another accident took place, when C-435 suffered engine problems near Comodoro Rivadavia and the pilot, Captain Justet, was forced to eject.

Between 1988-91 the Dagger, Finger, Mirage 5 and Mirage IIIEA/DA fleets received a new two-tone air superiority grey paint scheme, although C-401 and C-411 did not receive this, since they were retired from service. In 1989 flying hours reached 3,800, but budget cuts intensified after the takeover of President Carlos Menem. The Dagger/Finger fleet flew only 1,039 hours in 1991, with less than five operational aircraft on strength. When C-414 ingested an antenna during a low-level flight it was not repaired.

The cockpit of a Finger.
(Author's archive)

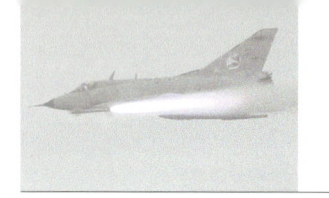

Latin American Mirages

A Finger drops a 500lb (227kg) FAS 300 cluster bomb with either 88 or 200 submunitions during weapons trials. (VI Brigada Aérea)

A Dagger wearing a rare paint scheme during Exercise Zonda 84 at Mendoza. (Author's archive)

By the early 1990s the FAA had begun to study the replacement of its combat jets, which now comprised the A-4B/C, Mirage III and the Mirage 5/Dagger/Finger. Both the F/A-18 Hornet and F-16 Fighting Falcon were assessed, but the US was unwilling to export either.

On 25 October 1993, C-427 suffered a fire during ground tests at ARMACUAR and was completely destroyed. This was followed by the loss of C-405 on 31 May 1994. Departing for a bombing training mission with live weapons, the pilot had to abort the take-off. Because of a misunderstanding with the control tower, the barrier was not erected, and the aircraft went off the runway and caught fire. Fortunately the pilot managed to escape safely.

By 1995 the budget had been slightly increased and Escuadrón I could put almost all available aircraft in service, with the exception of C-401, C-411 and C-414. In addition, C-408 was being used by the CEASO (Centro de Ensayos de Armamentos y Sistemas Operativos, Operational Systems and Weapons Test Centre), based at Río IV. In the same year the number of exercises was stepped up, but on 14 September 1995 another loss occurred, when Captain Raúl Gómez had to eject from C-413 after suffering porpoising.

Argentina

By January 1997, the Fingers and Mirage IIIs had been joined by the Mirage 5A Mara from Río Gallegos and they formed the Escuadrón X Cruz y Fierro (Instrucción), together with two-seat Daggers and the Mirage IIIDA.

International exercises

By the beginning of 1998, runway repairs at VI Brigada Aérea saw the Mirage IIIEA and Fingers deployed for more than three months to IV Brigada Aérea at Mendoza. On 14 August 1998, four Mirage IIIEAs and four Mirage 5s were deployed to V Brigada Aérea to operate with the OA/A-4AR Fightinghawk based there during the Águila 1 exercise, for which were joined by five F-16Cs and one F-16D of the 160th Fighter Squadron of the 187th Fighter Wing, Alabama Air National Guard. The Mirages performed air-to-air missions and the Argentine aircraft had the opportunity to fly mock combats against the F-16s.

The Águila II exercise took place at V Brigada Aérea, from 18–28 April 2001, and involved 4 Mirage IIIEAs, 4 Fingers, 2 Mirage IIIDAs, 2 Mirage 5A Mara, 14 OA/A-4ARs, plus 8 F-16Cs and F-16Ds of the 121st Fighter Squadron of the 103rd Fighter Wing, Columbia District Air National Guard.

In May 2002, the multinational CRUZEX exercise was conducted with the Chilean Air Force, the Armée de l'Air and the Brazilian Air Force, operating from Base Aérea de Canoas in Brazil. The Argentine contingent consisted of Finger and Dagger two-seaters, which operated alongside Armée de l'Air Mirage 2000C/Bs from EC 1/2 'Cigognes' and EC 2/2 'Côte d'Or', a KC-135FR Stratotanker and an E-3F Sentry. Brazil contributed examples of their F-5E, AMX and Mirage IIIEBR, and Chile sent Mirage 50 Pantera single- and two-seaters and a Boeing 707-330B Águila tanker.

Based on a scenario of one country invading another, CRUZEX included simulated attacks made by a multinational force. Each sortie included four Fingers, two Panteras, two F-5Es, two Mirage IIIEBRs and two or four Mirage 2000s, together with the AWACS and KC-135FR. Four AMX jets (two each for attack and reconnaissance) operated from other bases, as did Chilean and Brazilian tankers.

A major concentration of Latin American Mirages took place during CRUZEX 2002. Here, a Chilean Pantera, a French Mirage 2000C, an Argentine Finger and a Brazilian Mirage IIIEBR are seen at Canoas, Brazil, in May 2002. (Santiago Rivas)

Latin American Mirages

Fast and low, a Finger passes the V Brigada Aérea control tower during an airshow.
(Santiago Rivas)

The efficiency of the Argentine unit was evidenced by the fact that they were the only squadron to complete 100 per cent of planned sorties. The other nations were also impressed by the very proficient low-flying skills of the Argentine pilots.

The Salitre exercise was staged at the end of 2004, at Base Aérea Los Cóndores at Iquique, Chile. In this exercise, the Finger was joined by the Chilean F-5E Tigre III, Mirage 5 Elkan, Mirage 50 Pantera and Cessna A-37B, Brazilian F-5E Tiger IIs, and F-16Cs of the Air National Guard. Missions included attack, escort, air defence, air refuelling and search and rescue. All communications were conducted in English, and the idea was to adopt coordinated planning and doctrine to train for operations anywhere around the world under NATO procedure.

The Ceibo exercise of December 2005 saw participation by the Finger, Mara and Mirage III, together with the FAA's A-4AR, IA-63 Pampa, IA-58 Pucará and Paris. Non-Argentine assets included the Brazilian AMX, Chilean Mirage Elkan and Uruguayan Cessna A-37B.

A Finger lands at IV Brigada Aérea during a deployment to Mendoza.
(Santiago Rivas)

Argentina

A Dagger two-seater takes off on 10 August 1999. (Horacio Clariá)

The most significant operation conducted in recent years was the provision of air defence for the IV Cumbre de las Américas (Americas Meeting). Held during 4-5 November 2005, the event saw a meeting of 26 presidents from the Americas, including the President of the United States. Location for the summit was the city of Mar del Plata, 100km (62 miles) from Tandil. To prevent possible terrorist attacks, the Mirage IIIEAs operated armed with Magic missiles, and flew together with the A-4AR and the Pucará. The combat assets were controlled by a ground-based Westinghouse AN/TPS-43 radar and E-3 Sentry AWACS aircraft provided by the USAF. For the duration of the period that the presidents were in the city, FAA aircraft maintained standing two-aircraft patrols.

In 1998 another modernisation project was developed, with the intention of retaining the aircraft in service until 2015. Plans included the installation of A-4AR avionics, new weapons systems and electronics, and the addition of a flight-refuelling probe. This upgrade was never instigated, and by 2001 it had been superseded by plans to buy

A Finger lines up for take-off from VI Brigada Aérea at Tandil. (Horacio Clariá)

C-412, wearing kill markings signifying British ships, during the Ceibo 2005 exercise at Mendoza.
(Author's archive)

a number of ex-Qatari Air Force Mirage F.1EDA/DDA jets flown by Ala 14 of the Spanish Ejército del Aire, based at Albacete. The aircraft were inspected by FAA personnel, but the offer was not taken up. The FAA also inspected F-16A/B aircraft in storage at Davis-Monthan, with a view to upgrading these with Block 30 avionics, although this proposal also came to nought.

Continued budget problems meant that in 2010 the Fingers looked likely to remain in service for at least another five years. With this in mind, a contract was signed with the Segure Com company to provide a mission computer for the aircraft. The Argentine company Nostromo Consultora SRL was testing an example of the new computer in 2010, and preparing to install it in a Mirage 5.

A Finger armed with two 250kg (551lb) Expal bombs.
(VI Brigada Aérea)

Argentina

An air-to-air study of C-412.
(Santiago Rivas)

Thirty-two years after their arrival, the Fingers are now approaching the end of their career with VI Brigada Aérea. Their replacement is yet to be defined, although it could yet be the F-16 or the Mirage 2000. Whatever option is selected, it will represent the beginning of a new era for the FAA. In 2009, a new offer was received from Israel for second-hand Kfirs. (This was the third such offer, following previous offers in 1989 and 1997.) The latest offer concerned aircraft upgraded to Kfir C10 standard. Meanwhile, France offered ex-Royal Jordanian Air Force Mirage F.1s as a stopgap until Mirage 2000Cs could be delivered.

In 2010, VI Brigada Aérea consists of Escuadrón I with the Finger, the two-seat Dagger and the Mirage 5A Mara, and is tasked with fighter and attack missions. The

Two Fingers flying over northern Argentina.
(VI Brigada Aérea)

75

Latin American Mirages

A Finger takes off from Río Gallegos.
(Santiago Cortelezzi)

Israeli aircraft are equipped with Shafrir air-to-air missiles, while the French aircraft carry Magic missiles. The French-built jets can also carry a photographic pod with a KRB 8/24C camera for high-altitude and high-speed reconnaissance. For attack missions, the aircraft use general-purpose bombs. The Spanish-built Expal (Explosivos Alaveses) bombs are of the BK-BR type, and both 125 and 250kg (276lb and 551lb) classes are available. The 250kg bombs can be fitted with standard tails or parachute-retarded tails (BRP-S). Argentine bombs are also available, and these include FAS 300A and B cluster bombs and FAS 250 parachute-retarded bombs. The FAS 260 is intended for use against runways and bunkers (with the launcher containing nine

A Finger fires a Shafrir missile.
(VI Brigada Aérea)

A Finger under the snow at Río Gallegos in 2009.
(Santiago Cortelezzi)

or 18 bombs each of 37kg/82lb), and these are used together with the similar FAS 280 that carries high-fragmentation, parachute-retarded bombs of 34kg (75lb) each. The FAS 500 is a submunitions dispenser, while the 500lb (227kg) FAS 850 Dardo 1 is fitted with a rocket engine for increased penetration. The Dardo 2 is a standoff glide weapon with GPS guidance. The FAS 800A and B are of 125 and 250kg capacity respectively; they are similar to the US Mk 82 series, but carry submunitions containing small steel balls. Although the Daggers/Fingers are capable of carrying rocket launchers, the FAA does not use these weapons. The AS.30 missiles have also been withdrawn from use.

A Dagger in the night.
(Santiago Cortelezzi)

Latin American Mirages

Mirage 5A Mara

On 4 June 1982, the fall of Puerto Argentino/Port Stanley was imminent and the conflict between Argentina and the UK over the Malvinas/Falkland Islands was close to an end. Together with Argentina's other armed forces, the FAA was making a valiant effort to change the course of a war fought against one of the most powerful countries in the world. Events at VI Brigada Aérea, at Tandil in Buenos Aires province, far from the theatre of operations, demonstrated Peru's allegiance in the war. In fact, with the exception of Chile, the major South American powers all supported Argentina during the conflict: Bolivia, Brazil, Colombia and Venezuela also assisted Argentina, sending ammunition including bombs and rockets, as well as spares and many other items.

Ten Mirage 5P fighters from the Fuerza Aérea del Perú (FAP, Peruvian Air Force) landed at VI Brigada Aérea on 4 June, flown by Peruvian pilots. Their intention was to continue south, and the Peruvians also offered volunteer pilots to fight against the British forces. The Argentines declined this offer but the aircraft were immediately pressed into service, and assumed the serial numbers of aircraft lost in combat: C-403, C-404, C-407, C-409, C-410, C-419, C-428, C-430, C-433 and C-436. A number of pilots at Tandil, together with others returning from their deployment bases, were quickly briefed on the differences of the Mirage 5P compared to the Daggers in service with the FAA. Pilots were trained in the use of the AS.30 air-to-surface missile, also delivered by Peru.

The aircraft were in fact purchased under the terms of a contract signed in 1981. Negotiations for the transfer of the jets began in 1978, when a war with Chile seemed imminent. The plan was to deliver the Mirages some time later, but because of the Falklands War, they were prepared as quickly as possible and sent to Argentina. The

A Mirage 5P shortly after its arrival from Peru. (Author's archive)

Argentina

Mirage 5 C-404 during its early period of service with the Argentine Air Force. (Author's archive)

flight to Tandil involved flying over Bolivia at very high altitude and in complete radio silence. After a stopover at Jujuy, they continued to Tandil.

To improve the capabilities of the aircraft, the Ambar contract was signed in September with a US-based company, covering the provision of IFF equipment. By December, five aircraft had received IFF equipment, followed by the remainder in 1983.

Since the Mirages' electronics were old, it was decided to upgrade them, using some technologies from the Finger project, which had been developed to improve the capabilities of the Dagger. Meanwhile, the former Peruvian aircraft were integrated within VI Brigada Aérea, together with Dagger two-seaters that were used for training missions. Between 1982 and 1984 the Mirage 5Ps were deployed many times from Tandil, conducting exercises with the Mirage IIIC.

Two Mirage 5Ps with X Brigada Aérea in Río Gallegos. (Author's archive)

Latin American Mirages

Very low flying over Río Gallegos.
(Author's archive)

By the end of 1985, Escuadrón X de Caza 'Cruz y Fierro', part of X Brigada Aérea based at Río Gallegos, Santa Cruz province, had submitted a request to exchange their five Mirage IIICs for Mirage 5Ps. Their annual report stated that 'considering the operational responsibilities of the tasking imposed on the Brigade, it is apparent that with the equipment assigned (Mirage IIIC), the unit does not meet the operational standards in terms of quantity and quality, so replacement by more modern material (Mirage 5P) is requested, in order to accomplish the assigned task'. This request was considered by the head of the FAA and the order was given to send the 10 Mirage 5Ps to X Brigada Aérea in 1986.

On 7 November 1986, the Mirage IIIC fighters at Río Gallegos were sent to Escuadrón 55 at Mendoza. Six days later, the first five Mirage 5Ps arrived at the base, followed by the remaining five during 1986-87. In 1987 a total of 1,020 flying hours were assigned to the unit and a total of eight aircraft were operational. Meanwhile, the FAS-430 Mara project was developed by the Dirección General de Sistemas, following a request from Escuadrón X to improve the capabilities of the fighters.

Mirage 5 upgrade

The FAA decided to bring its Mirage 5Ps to a level similar to that of the Dagger (now being transformed into the Finger). Upgrades would address the armament control

Of these eight Mirage 5s, five have already been modified to Mirage 5A Mara standard, while three remain as Mirage 5Ps.
(Author's archive)

Argentina

A Mara flying over the Pampas, following the move to Tandil.
(VI Brigada Aérea)

box, the missile/gun control box, the firing programmer, the intervalometer and other systems. The upgrade would be carried out by the local Aerocuar company, together with the FAA, at the Área de Material Río IV.

A Canadian Marconi computer was also added for autonomous navigation, together with a Doppler radar, an Omega navigation system, an RWR developed by Aerocuar, a radio altimeter, DME, and a new artificial horizon and gunnery sight. The forward fuselage was changed, the Dagger nose and RWR antennas were added. The weapons pylons and the central gyroscope from the Dagger were installed and it was planned to install an Elta radar, but this was never realised. The upgraded aircraft were named Mirage 5A Mara (after a Patagonian hare) and the serial numbers were changed from the C-4XX series to the C-6XX series.

C-630 was the prototype Mara and was used for navigation tests, which revealed deficiencies with the RWR system. After these had been addressed, all the aircraft were modified between 1987-91, and then returned to Río Gallegos. Back at their base, the

Flying close to Chaitén Volcano on the southern border with Chile.
(VI Brigada Aérea)

Maras prepare for a four-ship formation take-off. (Horacio Clariá)

personnel of Área de Material Río IV installed the 97J gunsight. In 1988 the CES-3 and ADP4 adapters were added, permitting carriage of the Matra R.550 Magic missile. This work was undertaken by private contractors.

The zone around Río Gallegos was one of the 'hottest' in the country, thanks to its proximity to the Malvinas/Falklands Islands and southern Chile, and this meant the unit had a busy operational schedule. Each day, aircraft from the unit flew close to the border with Chile, and over the South Atlantic, into the British-declared exclusion zone around the islands. Although no visual contact was established with the RAF Phantoms flying from RAF Mount Pleasant, or the Chilean Mirage 50s from Punta Arenas, sometimes the fighters came very close to an engagement.

Starting the engine. (Author's archive)

Argentina

Chaff and flare launchers on a Mara.
(VI Brigada Aérea)

A Mara takes off during the Ceibo exercise, providing another view of the chaff and flare launchers.
(Chris Lofting)

X Brigada was considered the elite unit of the FAA, and each pilot had to have at least 700 hours on the Mirage in order to be accepted. The unit could not train its own pilots, since they did not possess two-seaters, and because the weather in the region made training very difficult.

In 1989 a total of 1,650 flying hours were assigned to the unit and four aircraft were delivered by the Río IV workshops. A first loss took place on 13 March, when C-607 suffered an engine failure. The aircraft was lost and the pilot ejected safely. The related SAR missions were conducted by Aero Commander 500U T-132 of the unit's Escuadrilla de Servicios, and an Army Bell UH-1H, supported by a Westinghouse AN/TPS-43E radar of Grupo 2 de Vigilancia y Control del Espacio Aéreo.

In January 1991 the unit received C-633, the final Mara to be delivered. By then, budget cuts in the FAA had begun, and the Mara program was delayed by almost two years. Other problems were the lack of air-to-air missiles and the fact that the RWR was no longer up-to-date. With the exception of C-619, the aircraft also lacked chaff/flare dispensers.

In 1993, only 1,050 hours were assigned to the unit and, despite the worsening budgetary situation, amplifiers were installed on the RWR. Only C-628 was fitted with the revised RWR and amplifiers in service, while C-630 was the only operational aircraft to receive new chaff and flare dispensers, but their cartridges were different to the others, since this aircraft served as the prototype. The brigade also lacked Magic missiles. Despite this, the aircraft began to be used for air-to-surface missions and photo-reconnaissance. The Mirages tested a photographic pod containing a KRB 8/24C camera intended for use at high altitude and high speed.

For bombing practice, the Army range of Comandante Luis Piedrabuena was used. Bombing practice began in November 1993, the jets dropping 19 500lb (227kg) BRP bombs, achieving 90 per cent effectiveness.

In January 1997, the FAA issued order 01/97 calling for the transfer of C-604, C-610, C-619 and C-628 to VI Brigada Aérea, while C-603, C-609, C-630, C-633 and C-636 were sent to Río IV for an overhaul before going to Tandil. The aircraft went to their new base in the course of the year and with them the Grupo Aéreo X, now called 'Los Guerreros del Hielo' (The Ice Warriors), disappeared. The reasons for the disbanding of the unit were not only budgetary: now that relations with Chile and the UK had improved, the threat to Patagonia was much reduced.

For a year the Mirage 5A operated with the Escuadrón Instrucción (Training Squadron) of VI Brigada Aérea. Then, on 4 June 1998, 16 years after the first arrival of the aircraft, the Escuadrón X 'Cruz y Fierro' (Instrucción) was re-established, but now as part of VI Brigada Aérea at Tandil. The unit also operated the Mirage IIIDA and the two-seat Daggers. Although it continued the tradition of 'Los Guerreros del Hielo', the unit's task was to train new Mirage pilots arriving from the Escuela de Caza (Fighter School) at Mendoza. This squadron was created in response to an original proposal in the 1980s that envisaged the formation of Mirage fighter school, which would have flown the Mirage IIIC and the two-seaters from Mendoza.

One week after the creation of the new Escuadrón X, C-604 was lost when one of the main landing gear legs failed. Major Luis Briatore ejected safely. On 8 August 2000, another accident took place, while C-609 was performing a training flight together with Mirage IIIDA I-002. The Mirage 5 pilot, Major Vincenzo Sicuso, was a pilot of the Aeronautica Militare Italiana (Italian Air Force) on exchange with the FAA. He explained that *'a bird got into the engine and caused it to stop. Eleven seconds had passed since I was airborne and I tried to start the engine again two times, without success. Looking ahead I saw the Federación Hill in front of me, so I ejected immediately'.* Less than three seconds after the ejection the Mirage hit the ground and the pilot recovered safely close to the crash site, where he was rescued.

The Mara today

Twenty-seven years after their arrival, the Maras are still in service with Escuadrón I, together with the Fingers and Daggers of VI Brigada Aérea, and the Mirage IIIEA/DA

A Mara in the snow in 2009. (Santiago Cortelezzi)

jets of Escuadrón II. The Maras continue to be used to train new pilots for the Mirage. Beginning in 2004 they received a new light grey paint scheme, similar to that used by the remaining Mirage/Dagger/Finger fleet. The Maras also received new chaff/flare dispensers, similar to those found on the Kfir. These are installed on the lower part of the fuselage, close to the engine exhaust, with four launchers for 20 cartridges on each aircraft.

Ceibo 2005 is the only multinational exercise in which the Maras have taken part. From 12-27 November 2005, the air forces of Argentina, Chile, Brazil and Uruguay met at Mendoza and conducted air-to-air and air-to-surface training.

Despite its age, the Mara will not be upgraded and in 2010 the type was awaiting replacement, while the quantity of operational aircraft was slowly decreasing. The original plan was to withdraw the Mara from service by 2005, but with no replacement in sight, the decision was taken to extend the type's service. Ultimately, the FAA intended to buy Mirage 2000C/B fighters to replace the entire Mirage fleet.

Mirage IIIB/C

FAA Mirage IIICJs in Israeli service

Israeli serial numbers
Since the FAA did not document which serial numbers were issued to which particular Israeli registrations, it is not possible to precisely match airframes. The Israeli serial numbers of Mirage IIIs delivered to Argentina were as follows: 03, 07, 29, 30, 33, 47, 48, 50, 51, 53, 58 (C-714), 59 (C-713), 66, 71, 76, 77, 79, 80, 98 (C-711), 86 (C-720), 87 (C-722) and 89 (C-721). These serial numbers were applied by Israel and are based on the original Dassault construction numbers. For example, in Israel, Mirage IIICJ 6603 became Mirage IIICJ 03, 6607 became 47 and so on.

For example, when C-713 arrived at the museum in Hatzerim in mid-2003, it was initially misidentified by the IDF/AF as 158 (C-714), an aircraft with 12 victories (plus another unconfirmed). Subsequently, curator of the museum Tsahi Ben-Ami was able to uncover hidden identification that confirmed this aircraft was actually serial number 159.

Mirage III victories in IDF/AF service
Almost all of the other Mirages exported to Argentina were credited with air-to-air victories during their service with the IDF/AF. C-713, which carried the Israeli serial number 59 (as well as 159, 259 and 459) was the Mirage credited with most kills in the IDF/AF. In total, the aircraft claimed 13 victories during the Six-Day War, the War of Attrition, and the Yom Kippur/Ramadan War. The aircraft was seriously damaged in a take-off accident on 7 October 1973 and was rebuilt by IAI, returning to service in 1974. It was subsequently used for reconnaissance with the serial number 459, and carried the six different locally built noses for the reconnaissance mission. Later, it once again received serial number 159.

For example, serial number 133 destroyed one Egyptian MiG-17 and one MiG-21 and a Syrian MiG-21. C-720 (previously 186) claimed three MiG-21s shot down on 24 October 1973. C-722 (previously 787) shot down an Egyptian MiG-21 over Tel Aviv. C-714 (previously 158, and also marked as 758 and 458) had 12 victories and one probable. When cross-examined against – meanwhile available – Arab documentation, almost 50 per cent of these claims proved exaggerated! For details, see 'Arab MiGs, Volume 2', also to be published by Harpia Publishing.

Dassault exported the Mirage IIIC to Israel as the Mirage IIICJ. Since related Argentine documents and aircrew refer to the aircraft as the Mirage IIIC, this designation is used in the following chapter.

Mirage IIIC with serial numbers 59, 159, 259 and 459 with 101 Squadron, IDF/AF of the Heyl Ha'Avir. Serves as C-173 within FAA. Claimed victories by Israel are mentioned in list below.

Aircraft/missile	Operator	Date
MiG-21	Syria	14 07 66
Il-14	Egypt	05 06 67
MiG-19	Egypt	06 06 67
MiG-21	Egypt	07 07 69
MiG-21	Egypt	20 07 69
Su-7	Egypt	11 09 69
MiG-21	Egypt	11 11 69

Aircraft/missile	Operator	Date
MiG-21	Egypt	04 01 70
MiG-21	Egypt	06 03 70
MiG-21	Egypt	10 07 70
MiG-21	Egypt	10 07 70
MiG-21	Syria	13 09 73
AS-5	Egypt	06 10 73

At beginning of June 1982, British troops were advancing on Puerto Argentino/Port Stanley in the Malvinas/Falkland Islands, and surrender was drawing close. While this was happening, negotiations had begun in Israel concerning Argentine acquisition of a batch of 23 Mirage IIICJ, BJ and RJ aircraft that had been withdrawn from service. Locally named Shahak, these aircraft had seen considerable combat during the Six-Day War and Yom Kippur War, claiming many victories against Arab aircraft. Among them, the Mirage 159 (which became C-713 in Argentina) claimed to have 13 victories to its name. Another example, 158 (later C-714) had 12 victories and one unconfirmed. One Mirage IIICJ and one Mirage IIIBJ had been tested with canard foreplanes, and Mirage IIIBJ 988 was for a short time equipped with a J79 engine during Kfir tests, although this aircraft did not go to Argentina.

Despite many claims that Argentina did not fully replace the aircraft lost during the war, the 34 combat jets lost in the conflict (11 Daggers, 10 A-4Bs, 9 A-4Cs, 2 Mirage IIIEAs and 2 Canberras) were replaced by a similar number of aircraft (23 Mirage IIIB/Cs, 10 Mirage 5Ps and 1 Mirage IIIBE), while Pucará losses were replaced by newly built aircraft.

After the negotiations, 19 Mirage IIICJs (registered as C-701 to C-719, and including a Mirage IIIRJ, modified to Mirage IIIC standard, which became C-711) and three Mirage IIIBJs (C-720 to C-722) were acquired. These aircraft were drawn from a total of

Comodoro Sapolski after his first solo flight in a Mirage IIICJ in Israel. He was the first Argentine pilot to solo on the type.
(Vicecomodoro Maggi)

Due to the embargo, the Mirages were painted with Peruvian markings for the trip to Argentina. They are seen here in the IAI hangar at Tel Aviv airport, prior to shipping. (Author's archive)

70 Mirage IIICJs, 4 Mirage IIIBJs and 2 Mirage IIIRJs that Israel had originally bought from France. Captain Guillermo Ballesteros was sent to Israel to conduct flight tests, which demonstrated that the aircraft were satisfactory. Twenty-two aircraft were sent by ship to Buenos Aires. The Mirages were painted with Peruvian markings to circumvent the embargo against Argentina. The 23rd aircraft remained in Israel to carry out tests associated with an aerial refuelling probe, but this project was later cancelled. Transported by C-130, this last aircraft arrived at Area de Material Río Cuarto (ARMACUAR) on 21 October 1983.

Preparations for service

On 18 December, ELMA *General San Martín* arrived at Buenos Aires with C-701, C-702, C-703, C-705, C-708, C-720 and C-722. These were taken over land to the Aeroparque Jorge Newbery airport. Comodoro Carlos E. Perona remembers that *'inside the Austral Airlines hangar the brown protection for the seawater was removed, the aircraft were armed, charged with hydraulic fluid, oil and other fluids. The engine, landing gear and all the systems were tested to check that everything was OK, and the aircraft were taken to ARMACUAR at Córdoba to be prepared for service'*. Captains Perona and Ballesteros were selected for the flight to Río IV, and the first delivery flight was made on 21 December by C-701 and C-722. Other pilots involved in the delivery flights included 1st Lieutenants Juan Carlos Sapolski and Selles. On three occasions three aircraft were delivered, instead of two. The second ship involved in the transport of the Mirages was ELMA *Tucumán*, which arrived on 1 February 1983 carrying C-704, C-706, C-707, C-709, C-713, C-717 and C-721. The third shipment was made by ELMA *General San Martín*, which brought C-710, C-711, C-712, C-714, C-716, C-718 and C-719 to Argentina on 29 March.

Initially, the FAA had discussed the possibility of using the Mirages for spare parts. At Río IV the aircraft were inspected, and some showed signs of having received combat damage. Other problems included sand within the structure, cables that were in poor condition, and some weak rivets. However, the structure was generally in good condition. The decision was taken to repair the aircraft completely, and put them into service. Work began immediately, but took a lot of time. Flight tests began in 1983, and involved 1st Lieutenant Mario A. Callejo and other pilots. Callejo confirmed that the main problem was the cables, with numerous failures during test flights. The HF equipment was changed and the aircraft received VOR, ILS, ADF, DME and two VHF radios. Although they had Cyrano I Bis radar, this would never be used.

The FAA planned to form two squadrons using the ex-Israeli Mirages, one at IV Brigada Aérea at Mendoza and another at BAM Gallegos at Río Gallegos, in the extreme south of the country. To make this possible, in 1983 a number of mechanics were sent from Mendoza to VI Brigada Aérea at Tandil, home of the FAA's Dagger operations, to learn about the Mirage and the Atar 9C engine.

Mirage IIIC over Argentina

On 26 February 1984, Escuadrón X de Caza ('Cruz y Fierro', but then known as the 'Guerreros del Hielo', or Ice Warriors) was established as part of Grupo 10 de Caza. At the same time, X Brigada Aérea was created at BAM Río Gallegos, with Comodoro Manuel Mir as the commander. In 1984 personnel from this squadron went to Río IV to test the new Mirages, and on 16 March C-702, C-703, C-704, C-706 and C-707 flew with 1,300-litre external tanks to Río Gallegos, with a refuelling stopover at Trelew. The pilots involved were Major Kajihara and 1st Lieutenant Marcelo Puig (who arrived at 16.15), and Captain Higinio Robles and 1st Lieutenants Carlos Perona and Daniel Galvez (landing at 17.15). On 19 March the brigade was officially inaugurated with a flypast by five Mirage IIICs.

Training began immediately, with missions over Patagonia, the Andes, the South Atlantic and Tierra del Fuego. As well as facing bad weather, the brigade was operating in the 'hottest' geopolitical zone in Argentina, with the British in the Malvinas/

The five Mirage IIICJs arrive at Escuadrón X de Caza at Río Gallegos on 16 March 1984. (Vicecomodoro Maggi)

Falklands Islands, and Chile on the other side of the Andes. Relations with Chile had been poor since the 1978 crisis, and had been compounded by Chilean support for the British during the war in the South Atlantic. For this reason, it was common to see the Mirages flying high over the Andes and the continental ice, crossing the channels of Tierra del Fuego at very low level, or patrolling the South Atlantic at very high altitude. The training flights served to show that Argentina had no intention of giving up its presence in the zone. Meanwhile, gunnery and bombing exercises were conducted, first using practice bombs on the base's firing range and later real bombs at the Army's Comandante Luis Piedrabuena range.

While Escuadrón X began its life in the cold of Patagonia, in February 1984 Escuadrón 55 was created at Grupo 4 de Caza of IV Brigada Aérea. The unit was named in honour of the 55 men of the FAA killed in the 1982 war. This squadron operated in the brigade together with surviving F-86F Sabres, the FMA/Morane-Saulnier MS.760 Paris and Aérospatiale SA.315B Lama helicopters. The chief of IV Brigade was Comodoro Claudio Correa, Grupo 4 commander was Comodoro Romeo Gallo, commander of the squadron was Captain Guillermo Donadille, and operations commander was Captain Norberto Dimeglio. The unit's other three pilots were Captains Yebra and Alberto Maggi and 1st Lieutenant Carlos Antonietti. All the selected pilots had experience on the Mirage family. Test flights began immediately at ARMACUAR and, as Comodoro Maggi remembers, the aircraft initially returned to base with all their failure panel lights illuminated.

Delivery of the aircraft was delayed, but on 17 October 1984, C-710 (Donadille), C-708 (Dimeglio), C-721 (Yebra) and C-722 (Maggi and Antonietti) took off from Río IV at 18.00 for the flight to Mendoza. C-722 was forced to return when the landing gear refused to retract. Arrival at Mendoza met with a curious coincidence. The date in question was the party-political day of General Peron's party, and celebrations were taking place in the city. The Mirage pilots, unaware of this, decided to make some low passes over the city. Their actions were construed by the so-called Peronists as a demonstration of FAA allegiance towards the party. In November a team returned to bring C-722 to Mendoza but it only managed the transit on the second attempt, on 28 November, shortly after the arrival of C-715 from Israel.

Once the squadron was operating the Mirages, eight new pilots began to receive instruction: Alferez José Luis Correa, Germán Demer, Jorge Luis Valdez, Gustavo Rodriguez, and Lieutenants Percy Juan Ryberg, Carlos Maroni, Luis Briatore and Raúl Estevez. In the course of the year, Escuadrón 55 flew 1,200 hours and Escuadrón X completed 800. For 1985, planned flying hours were reduced to 1,000 and 650, respectively, as a result of budget cuts.

Escuadrón X and the Mirage IIIC

Escuadrón X began 1985 with five Mirage IIICs, Aero Commander AC-500U T-144 and FMA IA-50 GII T-112, but on 28 April suffered the first accident with its new fighters. As Major Kajihara began his approach in C-707 and attempted to lower the undercarriage, one of the main units refused to deploy. The pilot took the aircraft over the base range and again tried to lower it, but without luck. Instead he ejected close to the base. The pilot was uninjured but the aircraft was destroyed.

A Mirage IIIC lands at Río Gallegos.
(Archive author)

On 30 July, during a flight by C-704 and C-706, the squadron recorded its first 1,000 flying hours. Soon after, C-712 arrived to replace C-707, and on 25 October C-706 suffered an accident when it overran the runway. The jet was repaired and flew again on 30 December. Although the squadron maintained a good level of serviceability with just five aircraft, the annual report for 1985 explained that 'considering the operational responsibilities of the tasking imposed on the Brigade, it is apparent that with the equipment assigned (Mirage IIIC), the unit doesn't meet the operational standards in terms of quantity and quality, so replacement by more modern material (Mirage 5P) is requested, in order to accomplish the assigned task'. The FAA command was in agreement, and ordered the transfer of 10 Mirage 5Ps to X Brigada.

In 1986 the Río Gallegos Mirages flew 550 hours, and on 29 July C-702 marked 1,500 flying hours for the unit. On 7 November the aircraft (C-702, C-703, C-704, C-706 and C-712) were transferred to Escuadrón 55. One of the Mirages had not been flown for a long time and was taken to Río IV by Captain Perona escorted by a Learjet 35A of II Brigada Aérea. Six days later, on 13 November, the first five Mirage 5Ps arrived, followed by the other five during 1986–87.

Servicing an aircraft with Escuadrón X.
(Archive author)

Latin American Mirages

Escuadrón 55

When Escuadrón 55 was formed, the FAA included six Mirage squadrons: one at Río Gallegos, one at Mendoza, two at Tandil and two at José C. Paz, near Buenos Aires. Because of problems with the training syllabus, each unit had to conduct its own training. Plans had been made to establish a Mirage school at Mendoza, to where all the two-seat Mirages would be transferred, and flown alongside the ex-Israeli Mirage IIICJs. In the event, this plan was never realised.

IV Brigada began 1985 with the arrival of the last two-seater, C-720, on 18 January. This was followed on 22 January by C-711 (the single Mirage IIIRJ, albeit fitted with a normal nose without cameras). C-709 followed on 28 January. Deliveries totalled 13 aircraft by the end of the year, with three more awaiting delivery (C-705, C-718 and C-719). On 22 January Alferez Correa and Valdéz made their first solo flights, and Lieutenants Ryberg and Maroni followed suit the next day. On 18 March Alférez Rodríguez and Demer soloed, followed on 19 March by Lieutenants Estevez and Briatore.

The FAA conducted its first dissimilar air combat (DACT) exercise in 1984. Operativo Zonda 84 featured participation by Grupo II de Vigilancia y Control del Espacio Aéreo and all brigades with the exception of I, II and X. During these exercises, which were repeated until 1990, simulated air combats were flown using various different combat types. In the course of these missions, the Mirage IIIC established a reputation as the most agile aircraft at high speeds. Comodoro Dimeglio remembers that *'the difference was made by having the same engine but with less weight and size than the other Mirages'*. Comodoro Perona agrees:

'It was a powerful and light aircraft, very docile, very good for air combat. It was lighter, with the centre of gravity further back, than the other aircraft'. In fact, the Mirage IIIC was longitudinally more unstable. With the centre of gravity near the tail, it had a tendency to put the nose up, and also offered a greater power-to-weight ratio than the other Mirages.

A Mirage IIIC is refuelled at Mendoza.
(Archive author)

Argentina

A formation landing at Mendoza. (Vicecomodoro Maggi)

By the end of the year the Mirage IIICs of IV Brigada Aérea had recorded 1,150 hours, a feat they would repeat in 1986. July 1985 saw negotiations to sell the aircraft to El Salvador, but they ultimately remained in Argentina. The DACT exercises continued in 1986 and the same year saw the arrival of the last three aircraft at Mendoza. The five aircraft from Río Gallegos arrived on 7 November, and Escuadrón 55 was left as the only unit equipped with the Mirage IIIC. On 12 August the F-86F Sabre was withdrawn from service, leaving the Mirage IIIC as the only combat type at Mendoza. Here, the Mirages operated together with the Paris, the Lamas, and the new FMA IA-63 Pampa jet trainers that were then arriving from the factory.

With the Mirages IIIs no longer in the best condition, C-701, C-702, C-703, C-704, C-706 and C-708 did not fly in 1986, and some of these jets would never fly again. Budgetary problems and the age of the fleet combined to demonstrate that the Mirage III was approaching the end of its career. 1986 also saw improvements made to air combat tactics. The aircraft were armed with two Shafrir air-to-air missiles and 30mm DEFA cannon, and generally carried a ventral 880-litre (194-Imp gal) fuel tank. For interceptions, the Mirages took off from their base, reached 15,000ft (4,572m) and then had reserves for five or six minutes of combat, in which they could reach 5.5g without losing speed. The Mirage then reduced power and landed with minimum fuel. Some

The original Israeli colours can be seen below the Argentine camouflage on this Mirage IIIC. (Vicecomodoro Maggi)

93

Latin American Mirages

C-715 received this two-tone light blue camouflage in 1987. (Suboficial Mayor Alfredo González)

pilots reported that the two-seater was more manoeuvrable than the single-seater, since it was more unstable. Despite carrying a second crewmember, the weight was the same for the Mirage IIIB, since the two-seater had the radar and various items of electronic equipment deleted.

By 1987 the Mirage III fleet had reached a total of 1,500 hours and seven aircraft were without their programa de vuelo (literally, flying programme, the annual hours assigned to each aircraft; if this flight programme was not issued, it was not planned to fly that aircraft during the year, and it could end up being cannibalised). Among the aircraft without a programa de vuelo were C-706, C-708, C-709, C-710 and C-717. In the same year, a new camouflage was tested on C-715, the green and dirty brown being replaced by a three-tone sea grey. This scheme was used for a short time before the aircraft was repainted in the original colours.

Between 7–11 August, 10 Mirage IIIs flew to Buenos Aires to take part in the flypast marking the 75th anniversary of the FAA. This would be the sole opportunity for the unit to participate in a flypast.

A formation of Mirage IIIC and Mirage IIIB jets. (Vicecomodoro Maggi)

94

Argentina

C-717, here with a special paint scheme during weapons tests, was the world's last operational Mirage III.
(Guillermo Galmarini)

The end

By 1988 there were 14 Mirage IIIs in service and seven without a programa de vuelo. All those in service had received Bendix DME-2030 equipment. On 29 July, during the reception flight after an inspection at Río IV, C-720 suffered an engine failure and crashed. The pilot was killed, but a mechanic in the rear seat managed to eject. C-718 suffered a minor accident when one of the main undercarriage tyres blew up, and the aircraft was brought to a stop by the runway barrier.

During weapons trials, Mirage IIIC C-717 was armed with a FAS 850 Dardo 2, an 800lb (363kg) GPS-guided standoff glide bomb.
(Author's archive)

Latin American Mirages

A line-up of different FAA Mirage types at VI Brigada Aérea, at Tandil (front to back): one Finger, two Mirage IIIEAs and four Maras. (Christian Amado)

Between 27 March and 7 April 1989, Operativo Gala 1 saw six Mirage IIIs deploy to Río Gallegos for air-to-air combat missions, which were flown between 30 March and 5 April. Another exercise, Operativo Ullúm, took place at San Juan on 6 October.

On 12 June, C-712 suffered an in-flight emergency. Although it was able to return to base, it was withdrawn from service shortly after. On 26 June, C-705 was lost near Media Agua, in San Juan province, claiming the life of 1st Lieutenant Carlos Bellini. C-704 and C-715 were out of service from 31 September, C-716 was repaired and C-710 was retired. By the end of the year eight aircraft were out of service, with nine left on the flight line. Of these, only six could be flown due to a lack of operational engines. The official IV Brigada history includes the following remark on the Mirage III: 'This weapon system showed its limitations for operations because of longitudinal instability problems, now investigated by the Accident Investigations Bureau.'

The last normal year of operations was 1990, during which the aircraft were slowly retired from service. By 1991 all had been removed from service and Escuadrón 55 was disbanded, although the aircraft were only officially retired in 1994. Thereafter, C-717 continued flying, now painted in a white, red and blue scheme, as part of the Centro de Ensayos de Armamentos y Sistemas Operativos (CEASO, Operational Systems and Weapons Test Centre) at Río IV. It was the final example of the Mirage IIIC operational with any air force and continued to fly until 2002. Both two-seaters remained with the Materiel Command at Río IV but they were never flown and were retired in 1998. Of the other aircraft, nine remain in storage at IV Brigada and the others were destined to become museum exhibits and monuments.

Argentina

Map of Argentina

Chapter 2

BRAZIL

Mirage IIIEBR/DBR

By the mid-1960s, the cutting edge of the Força Aérea Brasileira (FAB, Brazilian Air Force) comprised only the remaining Gloster Meteor F.Mk 8s of 1º Grupo de Aviação de Caça (1ºGAvCa) and 1º/14º Grupo de Aviação (1º/14ºGAv) and the Lockheed F-80C Shooting Stars of 1º/4º GAv. Both types had been conceived towards the end of World War II and were by now mainly used by the FAB for ground attack, with the result that Brazil had no genuine fighter aircraft.

The FAB requested funds to purchase new aircraft and develop a new air defence system. The proposal was approved by Brazilian President Castelo Branco, and in 1969 the Comissão de Estudos do Projeto da Aeronave de Interceptação (CEPAI, Interceptor Project Research Commission) was established to select the new fighter. The aircraft would be supersonic and radar-equipped and was to serve with a new unit to be established close to the new capital, Brasilia. Originally, the town of Luziânia was selected, but this location would have meant the jets would have to fly over populated areas during take-off and landing and would interfere with operations at Brasilia airport. The Aeronautics Ministry made a reconnaissance flight around the city until they found a suitable location, 10km (6 miles) from the town of Anápolis in Goiás State, and 150km (93 miles) from Brasilia.

After an initial study of the aircraft then being offered for export, the CEPAI pre-selected four: the McDonnell Douglas F-4C Phantom II, the Dassault Mirage IIIE, the BAC Lightning and the Northrop F-5A/B Freedom Fighter. The Phantom was preferred, on account of its payload capacity and the fact it was twin-engined, but the US refused to sell the F-4 to Latin America. The F-5 was also discarded since it was considered unable to fulfil the needs of the FAB (despite this initial rejection, Brazil later acquired large quantities of F-5B/E/F fighters).

Although the Lightning was not the preferred choice of the FAB, the Brazilian president initially opted for the British fighter. The Lightnings were to be accepted as payment to cover debts accrued by the UK for Brazilian coffee and cotton exports. The decision was taken in favour of the Lightning, but in August 1969, a matter of days before a contract could be signed, the president's ill health forced him to step down, to be replaced by General Emilio Garrastazu Médici. The decision to buy the Lightning was delayed as a result. Finally, the FAB's dissatisfaction with the Lightning saw it rejected, and the decision was taken to review the entire programme. The decision now lay with FAB, which decided on the Mirage IIIE. The Mirage was chosen since its performance was considered superior to the Phantom, and it had impressed in combat during the Six-Day War.

The initial plan envisaged not only equipping the unit at Anápolis with the Mirage, but also other fighter units. Early plans foresaw a purchase of up to 48 aircraft in different batches, but this was abandoned when the F-5E Tiger II became available at a very low price.

On 12 May 1970 a contract was signed with Dassault for 12 single-seaters, designated Mirage IIIEBR by Dassault and F-103E by the FAB (serial numbers 4910 to 4921). Also covered in the contract were four two-seat Mirage IIIDBR jets, known locally as the F-103D (serial numbers 4900 to 4903). The contract also included the training of aircrew and technicians.

While construction of the base at Anápolis continued, the initial Brazilian Mirage made a first flight at Bordeaux on 6 March 1972. On 23 May, the first team of eight FAB pilots left for France for training. All of them had more than 1,000 flying hours on combat aircraft. A number of technicians were also sent and the entire group was under the command of Coronel-Aviador Antônio Henrique Alves dos Santos, who had been selected to lead the FAB Mirage unit. Training flights were made using the Brazilian jets from the Armée de l'Air base at Dijon, and continued until the end of the year.

For those first pilots, conversion from the subsonic F-80 and Meteor to the supersonic Mirage was a major challenge. Thanks to its big delta wing, the Mirage was also harder to manage at very low speeds.

The organisation of the new unit would be completely different to that of other FAB formations. Both base and air unit were integrated into 1º Ala de Defesa Aérea (1º ALADA, 1st Air Defence Wing), reporting directly to the Comando de Defesa Aérea (COMDA) and part of the Sistema de Defesa Aérea e Controle de Tráfego Aéreo (SISDACTA, Air Traffic Control and Air Defence System). SISDACTA was Brazil's first national air control system and relied upon the Mirages to intercept any aircraft detected by ground radars and considered to be hostile. The system was officially inaugurated

A Brazilian Mirage IIIEBR during a pre-delivery flight in France. (Dassault)

on 23 October 1972 and was similar to that of Argentina, which combined the new Mirage IIIEA/DA with the radars of Grupo 1 de Vigilancia Aérea for the defence of Buenos Aires.

Training in France was completed by the end of September and the Mirages were prepared to be transported to Brazil. The first example, FAB 4910, was loaded into C-130E FAB 2456 of 1º Grupo de Transporte and arrived in Brazil on 1 October 1972, followed a week later by the second. The jets were reassembled from 16 October at the 1º ALADA base and were ready to begin operations by the beginning of 1973. On 27 March, Dassault test pilot Pierre Varraut made the first flight of a Mirage in Brazil in FAB 4910. Immediately after, the Brazilian pilots began flights. The new fighter was demonstrated to the Brazilian government and the public on 6 April, when a formation comprising FAB 4900, 4901, 4910, 4912, 4913 and 4914 flew over Brasilia. This initial presentation was followed by an eight-ship flypast over Base Aérea Santa Cruz, close to Rio de Janeiro, during Fighter Aviation Day on 20 April.

Mirage operations

The last of the 16 aircraft was delivered in May 1973 and the unit was declared operational. A pair of Mirages was maintained on constant alert, ready to defend Brasilia and the airspace between that capital and the cities of São Paulo and Río de Janeiro. The aircraft were armed with the Matra R.530 missile, with radar or infra-red guidance, as well as the two DEFA 552 30mm cannon. The Mirages were equipped with a Thompson-CSF Cyrano II mono-pulse radar and a Thompson-CSF Doppler radar. Despite their potential to undertake offensive missions with bombs and rockets, this capacity was not employed by the FAB until 1995, since there were sufficient aircraft of other types to undertake the ground-attack mission.

Pilots selected for service with 1º ALADA were drawn from other combat units and had considerable experience. The six-month training syllabus included flights in the two-seaters followed by conversion to the single-seater, and concluded with air-to-air engagements over the gunnery ranges of Natal or over the sea. Gunnery practice was undertaken using a target towed by another Mirage. Once assigned to the unit, pilots

A Mirage IIIEBR – local designation F-103E – during its early years of service. (Author's archive)

were baptised as 'Jaguares', using this as their callsign, together with an individual number. Throughout the Mirage III's career with the FAB, the number of 'Jaguares' reached almost 300.

The first loss occurred on 5 September 1974, when FAB 4921 crashed, the pilot ejecting safely. This was followed by the loss of FAB 4920 on 2 September 1975. In order to restore Mirage numbers after these losses, contract 05/COMAM/77 was signed in 1977 and covered four additional Mirage IIIEBR jets. These attrition replacements arrived in 1980 and carried the serial numbers FAB 4922 to FAB 4925. All four were second-hand Armée de l'Air aircraft. They would also replace FAB 4912, which was lost on 28 June 1979.

On 11 April 1979, Decree 004 brought about a change to the organisation of the FAB, and 1° ALADA became 1° Grupo de Defesa Aérea (1° GDA). At the same time, the associated aerodrome became Base Aérea de Anápolis (BAAN), with the new organisation effective from 19 April. Meanwhile, SISDACTA became the Centro Integrado de Defesa e Controle de Tráfego Aéreo (CINDACTA-1, Air Traffic Control and Defence Integrated Centre).

Accidents continued with the loss of two-seater FAB 4900 on 20 November 1980, claiming the life of Teniente Coronel Aviador Mauro Lazzarini de A. Silva, the first FAB pilot killed on the Mirage.

In the course of 1981, the Mirage's bare-metal finish was replaced by a two-tone grey air superiority scheme. The process was a gradual one, and some aircraft were only painted during modernisation some years later.

Ilyushin interception

On 2 April 1982, Argentine forces landed on the Malvinas/Falklands Islands, beginning a war with the UK over the possession of the islands. Brazil declared its support for Argentina, and put its bases on alert. Although close to the theatre of war, the possibility of taking part in combat operations was never studied. At 20.00 on 9 April, the CINDACTA-1 radar detected a large aircraft approaching from the north without authorisation. Air defence operators made radio contact and the pilot declared the aircraft to be a Cubana Ilyushin Il-62 flying to Buenos Aires, and carrying the Cuban ambassador to Argentina. The airliner was ordered to land at Brasilia but the pilot refused, and the order was given to scramble a pair of Mirages to intercept it. The pilots on alert were Major Aviador Paulo César Pereira in FAB 4916 and 1° Tenente-Aviador Eduardo José Pastorelo de Miranda in FAB 4922. Both were immediately sent to the pilots' room to prepare for take-off, while the aircraft were readied. The Oficial de Permanência Operacional (OPO, Operational On-Guard Officer) was Tenente-Aviador Roberto de Medeiros, who immediately contacted the Centro de Operações de Defesa Aérea (CODA, Air Defence Operations Centre) at Brasilia to ask for instructions.

While the aircraft were being prepared for take-off, a furious storm began and the base lost electrical power. The emergency electrical system was activated but the lighting between the taxiways and the alert shelters, and in the shelters themselves, was not connected to the system since these lights had only recently been installed. Take-off was delayed until the taxiways could be illuminated.

A rare photograph showing a Mirage IIIEBR flying together with a Meteor F.Mk 8, a T-33 and an EMB-326 Xavante. (Archive Pablo Kasseb)

Two Mirage IIIEBRs in formation. (Archive Aparecido Camazano Alamino)

The Operations Centre ordered the section to take off, using the 'Jaguar Negro' callsign. The controller used the callsign Thor. At 21.00 both aircraft taxied to the runway. Because of a problem with his navigation system, the wingman's take-off occurred a few minutes later. Meanwhile, the leader took off and was vectored to a height of 31,000ft (9,144m) heading for the intruder, who was flying over the city of Porto Nacional, to the north of Goiás State (now Tocantins). The wingman was held at 29,000ft (8,839m) and, because of the dense cloud layer, it was impossible to detect the intruder using the onboard radar. Instead, the Mirages were guided from the ground by Tenente-Coronel Anthony Blower of Núcleo do Comando de Defesa Aérea Brasileira (NuCOMDABRA, Brazilian Air Defence Command Centre). Minutes later, the leader was able to detect the target, which was now turning to the left. The Mirage climbed to 33,000ft (10,058m) and accelerated from Mach 0.98 to Mach 1.15, flying above the clouds. The pilot prepared the air-to-air weapons and approached the intruder, now followed very closely by his wingman. The Mirage pilots made visual contact and identified the intruder. The interceptors formated behind the airliner at a distance of two miles while the ground controller ordered the Ilyushin to land at Brasilia. The crew of the airliner responded that their destination was Ezeiza International Airport in Argentina. Since the airliner refused to land at Brasilia it was decided to make it known to the intruder that he had been intercepted. The Mirages were ordered to approach and make themselves visible to the Cuban airliner. The leader went to the right and the wingman to the left of the airliner, until both were close to the cockpit. While this was happening, Major José Orlando Bellon (Chief of Operations at CINDACTA-1) communicated with the Cuban pilot in English, saying *'Cubana 1225, you have been intercepted! There are two fighters alongside your aircraft. You are ordered to land in Brasilia immediately!'* This communication was followed by a silence from the pilot of the Il-62 before he answered *'Roger, Roger, Brasilia! Give me instructions'*. Ground control guided the Cubana aircraft until it landed in Brasilia, followed closely by the Mirages. The interceptors were short on fuel and the weather was still very bad, so as soon as the Cubana jet approached Brasilia airport, the Mirages returned to their base. It was later discovered that the Cuban ambassador was trying to mediate in the war before the US ambassador arrived in Argentina. Although authorisation to fly over Brazil was denied, the Cubans decided to press on regardless. After landing at Brasilia, the Il-62 was authorised to take-off again five hours later.

For the remainder of the war, the Mirages were kept on alert, but they were not scrambled again. Meanwhile the F-5E took part in the only other interception conducted by the FAB during the conflict, escorting an RAF Vulcan that made an emergency landing at Rio de Janeiro.

New deliveries and modernisation

To replace Mirage IIIDBR FAB 4900 and 4902, the latter being lost on 18 August 1981, contract 01/DIRMA/83 was signed in 1983, detailing the purchase of a pair of two-seaters. Former Armée de l'Air aircraft, the two jets were delivered in 1984 and carried the serial numbers FAB 4904 and FAB 4905.

This purchase was followed by contract 11/12/DIRMA/87, with the intention of increasing the fleet through the addition of two additional two-seaters (FAB 4906 and 4907) and four single-seaters (FAB 4926 to FAB 4929). These aircraft were delivered during 1988 and 1989 and the aircraft were valued at 18 million US Dollars for the single-seaters and 9 million US Dollars for the two-seaters.

On 29 September 1988 the FAB Mirages were called upon to conduct another live interception. On this occasion a Boeing 737-300 operated by VASP was hijacked. The hijacker, Raimundo Nonato, threatened to crash the airliner into the Brazilian government building and kill President José Sarney. The situation became tenser when the hijacker killed the co-pilot and forced the pilot to fly to Brasilia. The order was then given to scramble two Mirages to intercept the Boeing 737. The Mirages were ordered to escort the airliner and in case the situation became out of control, to shoot it down, to prevent even greater damage. Ultimately, control of the airliner was regained and after landing at the small airport of S. Genoveva in Goiania, the hijacker was captured.

On 29 April 1989, Formula 1 racing driver Ayrton Senna was invited to fly in Mirage FAB 4904, becoming one of the few civilians to fly the aircraft in Brazil. Later, the fighter received his name in honour of the flight.

A Brazilian Mirage displays the revised paint scheme. (Author's archive)

Brazil

In 1992 one Brazilian Mirage was tested with a refuelling probe, although the system was never adopted. (Author's archive)

A three-ship Mirage formation. (César Bombonato)

With the Mirages showing their age, a modernisation effort was planned, and in 1989 work began at the Parque de Material Aeronáutico de São Paulo (PAMA SP) and at Anápolis. Changes included a pressure fuel load system, new electronic equipment and installation of canard foreplanes, the latter to increase control at low altitude and low speed. The upgraded Mirages also received the capability to launch Matra R.550 Magic and Magic 2 missiles, although the latter was not available to the FAB. In total, six Mirage IIIEBRs and two Mirage IIIDBRs were modified locally, while the six aircraft purchased in France were modified prior to delivery. The paint scheme was revised again, with a lighter grey on the upper surfaces while the previous colour was retained on the lower surfaces.

Two Brazilian Mirages together with 30mm ammunition, an R.530 missile and target-towing equipment. (César Bombonato)

Mirage IIIEBR FAB 4914.
(Author's archive)

Based on the sheer size of Brazil, it was clear that it was impossible to provide complete air cover with the number of aircraft available. Matters were made more difficult by the fact that none of the available aircraft possessed adequate endurance to complete very long missions. As a result, it was decided to add aerial refuelling probes to FAB combat aircraft, modifying the entire fleet of F-5s and the new AMX. At the same time, the Boeing 707 transports were modified into KC-137 tankers. Mirage FAB 4929 was modified with a refuelling probe similar to that installed on the F-5 and on 22 April 1992 made the first air-to-air refuelling, from a KC-130H of 1º/1º GT. Although tests continued until 1993, and some pilots were trained in aerial refuelling, the system was not installed on the remaining Mirages. However, FAB 4929 retained the ability to be fitted with a refuelling probe until the end of its career.

New missions

On 12 April 1993, FAB 49154 conducted the first missile launch by a Brazilian Mirage, firing an R.530 over the Natal range. The ageing R.530 was retired soon after, and in 1995 the Mirage began training for the air-to-surface mission, alongside its primary role of interception. After a series of bombing, rocket and strafing exercises, the Mirages were declared as fully multi-role capable.

Another new development in 1993 was the first participation by FAB Mirages in an international exercise, during Operação Tigre II at Natal. The exercise saw Mirages in simulated air-to-air combat with USAF F-16C/D Fighting Falcons.

A Mirage in flight.
(César Bombonato)

Brazil

A pair of FAB Mirages head-on: an F-103E is followed by a two-seat F-103D.
(Johnson Barros)

Latin American Mirages

In 2002 FAB 4922 received a special paint scheme to celebrate 30 years of the Mirage in Brazil.
(Chris Lofting)

In 1997 the Mirages began to receive Rafael Python III air-to-air missiles from Israel. The same weapons pylons could accommodate the AIM-9B Sidewinder, also in the FAB inventory. At the same time, the Mirages also received chaff and flare dispensers.

In March 1997, Natal hosted Operação Mistral I, an air-to-air combat exercise that saw involvement by Mirage 2000Cs of the Armée de l'Air. The operation was repeated in March 1999 at Base Aérea de Santa Maria in the state of Río Grande do Sul.

By 1999, the end of the Mirage's career seemed close and the FAB launched its F-X programme, as part of Plan Fenix. The intention was to acquire a new-build combat aircraft for interception work, but also with an attack capability. Among the options were the Mirage 2000-5, the Sukhoi Su-35, the F-16C/D Block 50, the MiG-29 and the Saab Gripen, all of which were tested intensively.

Meanwhile, the last addition to the 'first-generation' Mirage fleet took place in 1999, when a pair of two-seat Mirage 5s was purchased from Zaire together with two single-seaters that arrived from the Armée de l'Air. The two-seaters were designated Mirage IIIDBR (F-103D) but they had two extra pylons under the fuselage, as found on the Mirage 5.

In July 2000, the F-X programme was officially launched and the different fighters on offer visited Brazil to be tested by the FAB. Among the pre-selected candidates were the Gripen, the Mirage 2000 and the Su-35. Although the Su-35 was initially preferred, the leading candidate emerged as the Mirage 2000, with Dassault offering the possibility of licence production by Embraer in Brazil. Furthermore, the introduction of the Mirage 2000 as a replacement for the Mirage III promised a straightforward transition.

A Mirage armed with Python III missiles.
(Author's archive)

Brazil

Carrying long-range tanks, F-103E FAB 4925 closes on a KC-137. Note the taped over squadron badge on the fin. (Johnson Barros/AirTeamImages)

Between 29 April and 11 May 2002, another joint exercise took place, Cruzeiro do Sul (CRUZEX I) including participation by Argentina with the Mirage 5 Finger, Chile with the Mirage 50 Pantera, the Armée de l'Air with the Mirage 2000C, E-3F Sentry and C-135FR, and the FAB with the AMX, F-5E, Tucano and Mirage IIIEBR. Operations were flown from Base Aérea de Canoas, at Porto Alegre, during which the Mirage III and 2000 provided air cover for the attack aircraft, with the AMX, F-5 and Tucano serving as aggressors. By then the locally designed Mectron MAA-1 Piranha air-to-air missile was being introduced to service, using the Python III pylons, and the Mirage fleet adopted the new weapon.

Control of Brazilian airspace was greatly improved by the arrival in 2003 of five Embraer EMB-145SA (named R-99A by the FAB) and three EMB-145RS (R-99B) aircraft. While the R-99A is equipped for airborne early warning and control, the R-99B is configured for electronic and signals intelligence missions. They entered service with 2º/6º GAv and were also based at Anápolis, operating alongside the Mirage. With the addition of the R-99A, the Brazilian Mirages became the first combat aircraft in Latin America to operate routinely with airborne early warning assets.

Mirages at Anápolis. (Author's archive)

Taking off.
(Chris Lofting)

Final years

While the F-X programme continued to evolve, it was decided that the Mirage III would have to be retired before the end of 2005, and the FAB began studying acquisition of a stopgap fighter until a definitive replacement could be purchased. Offers were received for ex-Israeli IAI Kfirs and Dutch F-16s, but they were not seriously considered.

During the final years of operations, a total of 15 Mirages were retained in service, with the same quantity of pilots, each of who completed 120 hours each year. Training was focused over the restricted airspace to the north of Anápolis, and was coordinated by the Brasilia Control Centre. In addition, one aircraft was kept on permanent five-minute minute alert in one of the shelters close to the runway end.

1º Grupo de Defesa Aérea was also equipped with two Neiva Regente and one Embraer C-95 Bandeirante for liaison, and three Embraer T-27 Tucanos for training. In addition, there was always a single Bell UH-1H detached from another unit, and responsible for SAR missions.

The last major exercises for the Mirage III took place in 2004, when they participated in Operação Terral XXXV, held in October at Base Aérea de Canoas (BACO). The exercise included air-to-air and air-to-surface components, using guns, bombs and rockets, and the Mirages operated together with the R-99A. In total, 10 Mirages and 3 Tucanos were deployed. The second edition of CRUZEX took place between 3 and 20 November, with Venezuela taking the place of Chile, and providing F-16A/B and Mirage 50EV fighters, while Argentina sent the A-4AR Fightinghawk and France once again deployed the Mirage 2000. The exercise took place in northeast Brazil, with the Mirages flying from Natal. CRUZEX was the major aviation exercise in Latin America, with around 70 aircraft and 1,200 personnel participating.

By 2005 the F-X programme was still not concluded, mainly as a result of political and economical problems. Brazil discarded the idea of a stopgap fighter, and examined the options of extending the life of the Mirage or of remaining without a manned interceptor. Finally, the decision was taken to buy a batch of 12 second-hand Mirage 2000C/B fighters from the Armée de l'Air.

The retirement of the FAB's Mirage IIIEBR/DBR fleet took place after the signing of the Mirage 2000 contract on 15 December 2005. However, the Mirage IIIs continued to fly until 31 December. In the course of 33 years of service they flew 67,000 hours. Pending the arrival of the Mirage 2000, the defence of Brasilia was entrusted to the F-5E.

Brazil

Mirage 2000C/B

Brazil acquired 10 former Armée de l'Air single-seat Mirage 2000Cs and a pair of two-seat Mirage 2000Bs for a total of 73.2 million Euros in line with a contract signed on 15 July 2005. The payment was made in six instalments from 2005 to 2010 and covered the aircraft, together with training and logistics. In service with the FAB, the Mirage 2000C is designated F-2000C, while the two-seater is known as the F-2000B.

After the aircraft had been prepared and the FAB crews trained during the first half of 2006, the first two Mirages 2000s were delivered to Brazil at Orange, France, on 10 August. The initial two aircraft to be handed over were single-seat FAB 4940 and two-seat FAB 4932. These both arrived at Anápolis on 4 September after a ferry flight from France. The ferry flight included a stop at Dakar, and saw the jets cover 9,000km (5,592 miles) in a total flying time of 11 hours and 20 minutes, including numerous aerial refuellings. On their arrival the new Mirages were displayed to the Brazilian president and other authorities and were then immediately introduced to service.

A Brazilian Mirage 2000B arrives in Brazil.
(FAB)

Mirage F-2000C FAB 4942 taking off in full afterburner.
(Johnson Barros)

111

Latin American Mirages

A Brazilian Mirage 2000C in flight.
(FAB)

Mirage F-2000C FAB 4942 receives fuel from FAB KC-137 2402 during exercise CRUZEX IV.
(Chris Lofting)

Brazil

Map of Brazil

CHILE

Mirage 50

Through its history, Chile has maintained a tense relationship with its three neighbours, Argentina, Bolivia and Peru. Chile has fought conflicts against both Bolivia and Peru, while on numerous occasions the country has almost gone to war with Argentina. Because of this, a near-constant arms race exists in the region, with the exception of Bolivia, whose scarce natural resources make it impossible to follow the lead of the other countries. From the 1940s, Argentina was in the lead in terms of its military, despite the fact that Chile received a significant quantity of material from the US after the World War II, including Republic P-47D Thunderbolt fighters, Douglas A-26 Invader attack aircraft and North American B-25 Mitchell bombers.

By the mid-1950s, the balance of power had shifted strongly against Chile. Chile was the last of the 'big three' to add jet aircraft to the inventory of its air force, when it obtained just six de Havilland Vampires in 1953. At the same time, Argentina operated 100 Gloster Meteors, while Peru introduced the F-80, the F-86 and the Hunter. When Chile received the Hawker Hunter in 1966, the disadvantage was reduced, but by the 1970s, Chile's adversaries had once again taken the advantage. Peru added 14 Mirage 5P/DP fighters to its air force, while Argentina boasted 50 A-4B Skyhawks and 12 Mirage IIIEA/DA fighters, and Argentina's Comando de Aviación Naval (COAN) acquired 16 Douglas A-4Q Skyhawks.

In the mid-1970s, the Fuerza Aérea de Chile (FACh) possessed no air-to-air missiles, while Argentina operated the Matra R.530 on its Mirages and the AIM-9B Sidewinder on COAN Skyhawks. In order to redress the balance, Chile bought a batch of Northrop F-5E Tiger IIs in 1976. However, this acquisition was not enough, as Peru increased its Mirage fleet and received the Sukhoi Su-22, while Argentina received 25 A-4C Skyhawks. Chile's Hunters were not capable of defeating the Mirages in aerial combat, and the fleet of FACh F-5Es was inadequate when compared against the array of air power possessed by the other countries in the region.

By the end of 1978 a border dispute with Argentina over the Beagle Channel zone, at the southern end of both countries, saw the FACh begin to deploy its forces. A war seemed inevitable. FACh inferiority against Argentina was evident, and was compounded by the fact that Argentina had received its first IAI Mirage 5 Daggers, equipped with Rafael Shafrir II air-to-air missiles. Furthermore, Peru had decided to participate in the confrontation and deployed a number of Cessna A-37B Dragonfly attack aircraft to Argentine bases. Chile had a force of only around 30 Hunters and 18 F-5s to face approximately 75 Mirages and 80 A-4s, plus assorted Su-22s, Canberras, F-86 Sabres and other types in service with Argentina and Peru.

Although the crisis was eventually resolved peacefully, the Chilean government understood that a significant effort was urgently needed in order offer some kind of resistance should a similar situation arise.

Driven by the success of the Mirage family around the world, and influenced by the fact that the fighter had been adopted by other countries in the region, the FACh began negotiations with Dassault for the purchase of a batch of the jets.

Mirage 50

France inherited 50 Mirage 5J fighters ordered by Israel after their delivery was cancelled in line with the weapons embargo declared after the Six-Day War. These were pressed into service with the Armée de l'Air as the Mirage 5F. In July 1979 the French government offered to sell eight of them, modified to Mirage 50CF standard. This promised a relatively rapid delivery, something that appealed to the FACh. The aircraft would receive new engines but would retain most of the remaining equipment, including the small Aïda fire control radar.

Before the end of 1979, a team of FACh pilots and mechanics went to France to undertake training with the Armée de l'Air. Pilots were schooled at Dijon, while mechanics went to Rochefort. The unit selected to operate the Mirage was Grupo de Aviación No. 4, part of I Brigada Aérea, which by then was based at Iquique, at Base Aérea Los Cóndores. Since 1976 the unit had been equipped with A-37B Dragonfly.

Initially the group would be based at Aeropuerto Arturo Merino Benítez, in the commune of Pudahuel, close to Santiago de Chile. The ultimate plan was to deploy the group to Base Aérea Carlos Ibáñez at Punta Arenas, on the southern tip of the country, very close to the zone where the crisis with Argentina had broken out.

Mirages in Chile

Once Dassault had modified the aircraft, on 26 June 1980 the first example was delivered to Chile. The final example arrived on 9 October. After being tested, the jets were loaded onto a ship and sent to Valparaíso, Chile's main harbour, 150km (93 miles) to the west of Santiago de Chile. From here they were sent by truck to their base and were assembled. After test flights in August, the Mirages were officially enlisted in Grupo No. 4 on 15 November 1980. The Mirages began operations around the city of Santiago while the base at Punta Arenas was being modified and upgraded to house the new unit. The Mirage represented a major technological advance over the Hunter, and required new infrastructure, especially at Punta Arenas, since this base had never previously housed a combat unit.

In Chile the Mirages received the serial numbers 501 to 508. While they flew their first training missions, Dassault built a batch of six Mirage 50C aircraft equipped with Cyrano IV-M3 radar, and two Mirage 50DC two-seaters. The Cyrano IV-M3 was an intermediate radar that came between those installed on the Mirage F.1 and the Mirage 2000, although some sources indicate that the aircraft in question actually received

Chilean Mirage 50CF serial number 505 just after delivery, and still wearing French camouflage.
(FACh via Álvaro Romero)

Agave radar. The two-seaters, despite their Mirage 50 designation, were fitted with the Atar 9C engine of the Mirage 5 and were identical to that version. Initially, the Chilean Mirages were equipped with Rafael Shafrir II missiles and were used to defend Santiago de Chile, a task previously performed by the ageing Hunter.

As Argentina and the UK fought for the control of the Malvinas/Falkland Islands between April and June 1982, the FACh Mirage 50s were used extensively to fly close to the border with Argentina. This forced the Argentine Air Force to maintain a constant watch on them, and to be prepared for a possible air strike by Chile. After the war, Chilean aircraft crossed the border and flew over Argentine airspace, while Argentine aircraft flew over Chile, but there were no encounters between the countries' aircraft.

Delivery of the newly built single-seaters began on 27 April 1982 and ended on 6 January 1983. These aircraft received the serial numbers 509 to 514. The two-seaters were delivered on 28 May 1982 (515) and on 24 November (516). The last of these was lost in an accident shortly after its arrival and was replaced on 1 June 1987 by an ex-Armée de l'Air Mirage IIIBE, which received the same serial number.

Mirage 50C FACh 509 with bombs and rockets. Note the different nose for the radar.
(FACh via Álvaro Romero)

Latin American Mirages

Mirage 50C FACh 511 over the Andes.
(FACh)

A Mirage 50CF with Cardoen CB250K and CB125K cluster bombs. Note that the serial numbers and national insignia have been covered with tape.
(FACh via Álvaro Romero)

A change of base

In order to keep the weapons system up to date, Programa Bracket began in November 1985. This included installation of canard foreplanes, two small strakes on the sides of the nose for vortex generation, and a radar warning receiver. These changes were made in all of the aircraft. Work was undertaken in Chile with the support of Israel Aircraft Industries (IAI), who also manufactured the canards. The first modified aircraft was 514, which began test flights in 1986.

Meanwhile, in March 1986 the aircraft were finally transferred to Base Aérea Carlos Ibáñez at Punta Arenas, from where they operated until the end of their operational career. Modernisation works resumed after the transfer, and were now undertaken as part of the Pantera project, which also included modernisation of the avionics.

At the same time, one of the aircraft was used to test the CB125K and CB250K cluster bombs, of 276lb and 551lb (125kg and 250kg) respectively, developed by the Chilean company Cardoen. The CB250K was added to the FACh inventory in small quantities and was exported to some countries, including Colombia.

Empresa Nacional de Aeronáutica (ENAER), part of the FACh, undertook the new modifications but all flight research was conducted by IAI. Again, serial number 514 was used as testbed and was sent to ENAER. Here the aircraft received the Elta EL/M-2001B Doppler-type fire control radar and the associated modern navigation and weapons system. The latter introduced the possibility of launching air-to-ground guided weapons. An EWPS-100 integrated electronic warfare suite was added, this being built in Chile by DTS. The EWPS-100 included the ENAER Caiquen III (DM/A-104) RWR, Eclipse (DM/A-202) chaff and flare dispensers, and a DM/A-401 electronic countermeasures system. The cockpit received a multifunction liquid-crystal display to the left of the main panel. On the central part of the panel was a gyroscope linked to a computer-

On 14 October 1988, FACh 514 was rolled out as the first aircraft modified to Pantera standard.
(FACh)

Latin American Mirages

FACh 514 during tests flights for the Pantera programme.
(FACh)

ised head-up display, working in conjunction with an Elop inertial navigation system. All armament, hydraulic and electric controls were upgraded and two extra hardpoints were added below the air intakes. The modifications were made using the same equipment that was used to upgrade the F-5E to Tigre III standard.

The first upgraded Mirage 50 Pantera was rolled out on 14 October 1988 at the ENAER facilities in Santiago de Chile. Besides the canards, the aircraft were outward-

On 21 March 1994 two Peruvian Mirage 2000Ps flew together with a Pantera during the FIDAE trade fair.
(Patrick Laureau)

Chile

A Mirage Pantera testing an IAI Griffin laser-guided bomb equipped with a 500lb (227kg) Mk 82 warhead. (FACh)

ly easily distinguished by their enlarged, Kfir-style nose, extended by almost 1m (3.3ft). Initially the 30mm DEFA guns were not fitted, but these were installed later.

Tests proceeded well and the modifications showed no signs of problems, and the order was given to modify other aircraft to the same standard. Meanwhile, on 18 June 1989, the second accident of the type took place, when serial number 512 suffered an onboard failure during take-off. The pilot ejected but the seat failed to work properly and he lost his life.

The second Pantera was delivered in 1990, but budget cuts since then meant the programme progressed very slowly. In 1993 the system was declared fully operational, but deliveries of modernised aircraft remained at a very low rate. The last example was finally delivered in 2002, when the two-seater 516 was handed over. In the 1990s the modified jets received a refuelling probe, Rafael Python III air-to-air missiles (replacing the Shafrir II) and Elta EL/L 8212 ECM pods. The latter could be installed on one of the forward fuselage pylons. French-built laser-guided bombs were tested but were not adopted by the FACh.

On 12 March 1993 another Mirage was lost close to Pecket, northwest of Punta Arenas. The pilot was forced to eject because of an undercarriage failure.

Pantera FACh 508 taxiing. (Álvaro Romero)

A Pantera with two F-16C
Fighting Falcons at Los Cerrillos
during FIDAE 92.
(Patrick Laureau)

Two Panteras refuel from Chile's
single Boeing 707 Aguila tanker.
(FACh)

A Pantera and an A-37B
Dragonfly refuel from the
Boeing 707 Aguila tanker.
(FACh)

122

Chile

Two Panteras accompany two A-37B Dragonflies over southern Chile.
(FACh)

FACh 513 in flight.
(FACh)

FACh 501 with an Elta EL/L 8212 jamming pod.
(Santiago Rivas)

Latin American Mirages

International exercises

During their career, the Mirage 50s took part in many exercises within Chile, but only once deployed to a foreign country for training. With the dawn of the new millennium, politics in the region changed, and tensions between all Latin American countries began to subside. In particular, Chile's relations with Argentina became more relaxed. Exercises took place in which the FACh participated alongside various other Latin American countries, with the notable exception of Peru.

The first of these exercises was Cruzeiro do Sul (CRUZEX 2002), held at the Brazilian Air Force base of Canoas, between 29 April and 11 May 2002. Chile sent three single-seat Mirage 50s and the two-seater 515, while Argentina sent IAI Mirage 5 Finger jets, the Armée de l'Air sent Mirage 2000C, E-3F and C-135FR aircraft, and the local Força Aérea Brasileira employed the AMX, F-5, Tucano and Mirage IIIEBR. Operations were conducted across southern Brazil, with the AMX, F-5 and Tucano acting as aggressors. During the deployment, the Mirage 50 flew together with the FACh Boeing 707 Águila tanker, making a direct flight from the north of Chile to the Brazilian base, flying over Argentina. This was the only time the Chilean Mirages left the country during their entire operational service. In 2003 Chile purchased five Dassault/Atlas Cheetah E fighters from South Africa. Serial numbers 819, 820, 827, 832 and 833 were used as a spares source for the Panteras. Although the FACh indicated its desire to purchase seven more aircraft (822, 823, 825, 828, 829, 831 and 834), this never took place. The airframe of serial number 833 is now in a derelict condition at the aviation museum in Los Cerrillos.

A Pantera taxies during the Salitre 2004 exercise in Antofagasta.
(FACh)

FACh 504 during the Salitre 2004 exercise.
(FACh)

124

FACh 503 undergoes maintenance. (FACh)

Between 25 September and 10 October 2004, the Salitre 2004 exercise took place at Base Aérea Cerro Moreno, at Antofagasta, and Los Cóndores, at Iquique. Salitre 2004 involved the air forces of Argentina, Brazil, Chile and the US Air National Guard. Argentina sent two Mirage IIIEAs, three Fingers, one Mirage IIIDA, one Lockheed L-100-30 Hercules and one Boeing 707; Brazil sent six F-5E Tiger IIs and one KC-137; Chile participated with eight F-5E Tigre IIIs, eight Mirage Elkans, six Panteras, eight A-37Bs, two CASA C-212-300s, one Sikorsky S-70A-39, two Bell UH-1Hs, one Learjet 35A and two Boeing 707s (the Águila tanker and the Cóndor ECM and airborne early warning platform). The US participation consisted of six F-16Cs supported by a KC-10.

Combined training missions took place from 27 September until 1 October, focusing on the integration of members of the different air forces. The operational stage began from 3 October, with combined and massive use of air power during air-to-air, air-to-ground and support mission scenarios. As an example of the missions, in the

FACh 503 during FIDAE 2006. (Katsuhiko Tokunaga/D.A.C.T.)

space of a few hours Los Cóndores was 'attacked' by FACh F-5s and A-37s that were repelled by the ground-based air defences and interceptors. After this attack and a simulated damage assessment, the coalition responded with a massive air strike using Panteras, Elkans and Fingers, escorted by F-16s and Mirage IIIEAs.

The Pantera remained in service at Punta Arenas, although it was planned to replace the fighter with F-16s. However, the region makes F-16 operations particularly complicated. A combination of strong winds and the particular soil characteristics mean the runways are frequently covered in stones, which is a particular danger to the F-16 with its low-slung air intake.

Eventually, tensions with Argentina having almost disappeared, it was decided to retire the Pantera without replacement. In December 2007, the fighter was withdrawn from service and the aircraft stored and offered for sale. The only country to show interest was Ecuador. Negotiations between the two countries took place during 2009, but had yet to reach a definitive outcome by September 2009.

FACh 515 over the Andes. (FACh)

Chile

Panteras 506 and 516 fly over the Strait of Magellan.
(FACh)

A Pantera two-seater flies low over the Strait of Magellan, accompanied by two single-seaters.
(FACh)

A two-seat Pantera banks at very low level near Punta Arenas.
(FACh)

Latin American Mirages

FACh 516 lands in 2007.
(Álvaro Romero)

FACh 504 touches down in 2007.
(Álvaro Romero)

The nose of FACh 505.
(Katsuhiko Tokunaga/D.A.C.T.)

128

Mirage 5 Elkan

By the early 1990s, the FACh had begun to study a replacement for its whole combat fleet. Under Proyecto Caza 2000, FACh combat assets would be unified, with a single new type replacing all the models then in service. Chile tested the F-16, the Saab Gripen, the Mirage 2000 and the Sukhoi Su-27/30 series, but this ambitious programme fell foul of budget cuts, and delays in acquisition of a new fighter were the result. Eventually, the plan of replacing all types was discarded and priority was given to replacing the older fighters in the inventory, and especially the obsolete Hawker Hunter.

Because it was impossible to find the budget to buy new aircraft, it was decided to purchase second-hand equipment to serve as a stopgap until a new generation of fighters could be introduced.

Among the types offered second-hand were a batch of IAI Kfirs, another of SEPECAT Jaguars, and one of Cheetahs from South Africa. However, the most interesting option proved to be 25 Mirage 5B fighters offered by Belgium, these being in good condition and available at an attractive price.

In 1989 the Belgian Air Force signed a contract with SABCA to modernise its fleet of Mirage IIIBA, BD and BR aircraft. All of these had been built in Belgium since the 1970s, and comprised single-seat, two-seat and reconnaissance versions, respectively. The original plan was to retain the Mirages in Belgian service until 2005 after implementing the MirSIP (Mirage System Improvement Programme). As a result, 15 examples were selected for upgrade, a quantity later reduced to 10 (BA04, BA11, BA46, BA60 and BA62, and BD01, BD03, BD13, BD14 and BD15). The jets were delivered to SABCA's workshops at Coxyde/Koksijde from 18 May 1989 and the first flight of a Mirage 5B MirSIP was made by BA60 in December 1992. The quantity to be modernised was later increased to the originally planned total, with another 10 single-seaters (BA01, BA23, BA37, BA39, BA48, BA50, BA56, BA57 and BA59).

An Elkan over northern Chile armed with Sidewinder training missile.
(FACh)

Latin American Mirages

FACh 718 still in Belgium in FACh camouflage with both Belgium and Chilean markings. (FACh)

FACh 703 near Antofagasta. (FACh)

MirSIP included the installation of a GEC Marconi head-up display with an Up-Front Control Panel (UFCP) to operate it, and a video recorder to monitor the HUD. A Weapons Delivery Navigation and Reconnaissance System (WDNRS) was added, working with a Thompson-CSF TMV 630 laser telemeter that worked in conjunction with a Honeywell radio altimeter, a SAGEM UTR 90 navigation and attack computer, and a SAGEM USLISS 9 INS – all operated via a Mil Std 1553B databus. Two multi-function displays were added to the cockpit and the IFF equipment was improved. Also installed were a single fuel pressure loading point and a liquid oxygen system, anti-collision lights, canard foreplanes and new wiring and cables. The airframe was subject to a complete overhaul, and new Martin-Baker Mk 10 zero-zero ejection seats were added.

With the disbanding of their operating unit (42 Squadron), it was decided to retire the entire Mirage 5 fleet by 1993, leaving the F-16 as Belgium's sole combat aircraft. Since the cost of cancelling the MirSIP program was so high, it was decided to continue with it and offer the aircraft for sale. The first modernised aircraft were delivered in November 1993 and then stored.

FACh 713 taking off for another mission during exercise CEIBO 2005. (Chris Lofting)

Chile

A Mirage 5BR seen from above. Note the missing canards. (FACh)

Latin American Mirages

On 19 July 1994 a contract was signed between Belgian defence minister Leo Delcroix and the FACh commander-in-chief, General Jaime Estay Riveros. The contract covered the purchase of the 15 modernised single-seaters and five two-seaters, plus a single two-seater and four Mirage 5BRs that had not undergone modernisation. The contract was worth 54 million US Dollars, to be paid in 18 months, and included an important batch of spares and a test bench for Atar 9C engines. The test bench was used by the Belgian Air Force at Bierset and was disassembled and sent by ship to Chile in June 1996. The aircraft were named Elkan (meaning 'Guardian' in the Mapudungun language used by the Mapuche Indians).

A further contract was signed, worth 1.5 million US Dollars, to provide training for eight Chilean pilots and a team of mechanics. In April 1994, a training unit was formed at Brustem/St Truiden air base, and was named Detachment MirSIP. This was equipped with three modernised two-seaters (BD03, BD04 and BD15) and two single-seaters (BA56 and BA62). Training continued until March 1995.

Elkan in Chile

Meanwhile, in Chile an investigation began into a supposed corruption case relating to the purchase of the aircraft. The process continued into early 2009, when the Chilean defence minister revealed that important documents related to the purchase had been destroyed some time previously.

In February 1995 the first five Mirage 5BAs were loaded into an Antonov An-124 that took them to Cerro Moreno air base at Antofagasta, where they were assembled and entered service with Grupo de Aviación No. 8 of I Brigada Aérea. The single-seaters received the serial numbers 701 to 715, the two-seaters became 716 to 720, the reconnaissance variants 721 to 724, and finally 725 was applied to the sole two-seater

Two-seater Mirage Elkan FACh 720 during exercise CEIBO 2005 in Argentina. (Chris Lofting)

that had not undergone upgrade. Shortly before their arrival, the last Hawker Hunters were retired from service, leaving Grupo No. 8 inactive.

The reconnaissance Mirages were used to replace the Hunters previously used in this role, but they had not received any overhaul or modernisation and were not in a very good condition. It was planned to send the non-modernised two-seater to Grupo No. 4, reinforcing the two trainers in use with that unit, and compensating for the fact that serial number 516 was for a long period at ENAER's facilities, where it was being converted to Pantera standard.

The Mirages arrived during 1995 and by 1996 the unit was declared operational. During an exercise on 7 November 1999, when two three-aircraft formations were preparing to land at Punta Arenas, one Elkan hit the ground heavily and the pilot ejected. The pilot escaped without injuries and the aircraft continued along the runway, stopping 200m (656ft) beyond the runway end, and sustaining heavy damage.

On 27 November 2001 another Elkan broke its nose gear while landing at Antofagasta. The aircraft suffered only light damage and was repaired. During that year it was decided to retire the non-modified aircraft, beginning with the two Mirage 5BRs, which were sent to the Escuela de Especialidades for ground training. One of them, serial number 722, was later delivered to the Museo Aeronáutico at Santiago de Chile. Serial number 725 was finally installed as a monument at the entrance of Cerro Moreno air base, painted in a two-tone grey scheme.

As recounted in the previous chapter, between 25 September and 10 October 2004, the Salitre 2004 exercise took place at Base Aérea Cerro Moreno in Argentina. The Elkans were active participants alongside aircraft from Argentina, Brazil, Chile and the US Air National Guard. This exercise was followed between 12-27 November 2005 by Ceibo 2005, hosted by two Argentine Air Force formations: IV Brigada Aérea in Mendoza, and V Brigada Aérea at Villa Reynolds, San Luis province. Ceibo 2005 saw participation by the air forces of Argentina, Brazil, Chile and Uruguay. Ceibo 2005 was similar to Salitre, with a coalition of aircraft from all countries fighting a fictitious enemy consisting of Argentine A-4AR Fightinghawks and Mirage IIIEAs, the adversaries also performing ground-attack missions. Argentina participated with one C-130B and one KC-130H, six IA-58A/D Pucarás, two IA-63 Pampas, four MS.760 Paris, two SA.315B Lamas, eight OA/A-4AR Fightinghawks, two Mirage 5A Maras, one Mirage IIIDA, six Mirage IIIEAs, four Mirage 5 Fingers, two Bell 212s and one Fokker F27. Brazil sent five AMX and one AMX-T, Uruguay three A-37Bs, and Chile sent five single-seat Elkans and one two-seater.

Mirage Elkan FACh 705 with an AIM-9 Sidewinder training missile.
(Chris Lofting)

Retirement

With the intention of expanding the Elkan's capabilities, in 2003 serial number 706 received a refuelling probe for test purposes. The probe was later installed on other examples of the Elkan.

After a long discussion concerning the acquisition of a new combat aircraft, in 2002 the FACh announced the selection of the Lockheed Martin F-16 as its new air superiority fighter. According to this decision, on 22 May 2003, Lockheed Martin announced the signing of a Foreign Military Sales (FMS) contract with the USAF worth 320 million US Dollars and covering construction of 10 F-16C/D Block 50 aircraft (6 single-seaters and

Latin American Mirages

4 two-seaters) for the FACh. On 23 June 2005 the first Chilean Fighting Falcon made its maiden flight and on 31 January 2006 the first two examples arrived at Santiago de Chile. The F-16s were put into service with Grupo de Aviación No. 3, which had been reactivated.

Meanwhile, by the end of 2005 the purchase of 18 F-16A/B Block 15 MLU aircraft (11 single-seaters and 7 two-seaters) had been agreed with the Royal Netherlands Air Force. The price was put at 180 million US Dollars, with deliveries taking place from August 2006 to September 2007. The former Dutch F-16s were delivered to Grupo de Aviación No. 8. With the arrival of the first 10 jets from the Netherlands, on 27 December 2006 the last 10 Elkans were retired during a ceremony at Grupo No. 10 at Santiago de Chile. Although the Elkans were offered for sale, no customer came forward.

Elkan FACh 714 lands at El Plumerillo in Argentina during Exercise CEIBO in November 2005.
(Chris Lofting)

Chile

Map of Chile

Chapter 4

COLOMBIA

Mirage 5COA/COR/COD

By the end of the 1960s, the Fuerza Aérea Colombiana (FAC, Colombian Air Force) combat fleet consisted of 6 Canadair Sabre Mk 4s, 4 North American F-86F Sabres, 16 Lockheed F-80C Shooting Stars and a quantity of Lockheed T-33As and AT-33As. As a result of their age, the F-80 and F-86 were retired by 1968, leaving the FAC without any genuine combat aircraft. This situation, combined with the activities of guerrillas inside the country, demonstrated the importance of introducing new combat aircraft, which were, at this time, not included among the military materiel provided by the US.

The refusal of the US to sell fighters meant that Colombia looked to Europe for a solution, and selected the Dassault Mirage, which offered both high performance and a relatively low price tag. In 1970 the FAC ordered a batch of 14 single-seat Mirage 5COA fighters, 2 two-seat Mirage 5COD trainers, and 2 Mirage 5COR reconnaissance aircraft. The 5COR variant was the first reconnaissance Mirage to be delivered to an operator in the region.

In 1971, FAC pilots and mechanics went to France to train on the new fighter, and by the end of the year the first examples of the Mirage had been delivered in France and transported to Colombia. In the early hours of 1 January 1972, C-130 FAC 1001 landed at Base Aérea Militar Germán Olano, at Palanquero, carrying Mirage 5 FAC 3024. Initial local tests of the first aircraft were planned for 20 and 21 March. During these tests, the first Mirages to fly in Colombian skies were FAC 3024, FAC 3021 and FAC 3025, which entered service with Escuadrón 212. Reporting to CACOM-1 (Comando Aéreo de Combate 1), Escuadrón 212 was activated on 25 March 1972. Deliveries of the single-seaters (FAC 3021 to FAC 3034) were completed on 17 July 1973.

Mirage 5COA FAC 3029 during tests in France. (Dassault)

On 13 and 20 March the two-seaters were delivered (FAC 3001 and FAC 3002), and the reconnaissance versions (FAC 3011 and FAC 3012) followed on 19 October and 19 November. The squadron was considered operational and began training with the new aircraft, but on 12 August the first attrition took place, when a Mirage 5COA was lost in an accident. Captain Guillermo Díaz Muñoz was killed when the Mirage suffered a type blow-out during take-off and the aircraft went off the runway.

In combat

Since the 1950s, Colombia has faced an almost permanent struggle against guerrillas, initially those sympathetic to the Liberal Party. Subsequently, the revolutionary guerrillas loyal to the Communist Party became the main threat, as their operations intensified. In 1964 the Fuerzas Armadas Revolucionarias de Colombia (FARC) and the Ejército de Liberación Nacional (ELN) were formed, these becoming the most significant guerrilla organisations. The groups' activities intensified in the 1970s and 1980s, especially as they became increasingly involved in drugs traffic and production. The FAC participated in the fight against the guerrillas from the outset, initially using the T-33 and the Cessna T-37C. Later, the Mirage 5 was also widely used.

In August 1987 the Colombian Navy sent the frigate ARC *Caldas* to the Gulf of Venezuela to establish a presence on three small islands that were the subject of dispute and were occupied by Venezuela. The Venezuelan Navy immediately sent the frigates ARV *Mariscal Sucre*, *General Urdaneta* and *Almirante Brion* to the zone in order to force the Colombian vessel to leave. The Colombian government decided to send Mirage 5s to over-fly the zone and make passes over the Venezuelan ships, threatening to attack them. Venezuela's air power was superior, and its air force deployed F-16s to the area to counter the presence of the Mirage 5 and to attack the ARC *Caldas*. Finally, three days after the standoff began, the Colombians withdrew from the area and the crisis ended.

This incident in the Gulf of Venezuela served to demonstrate the inferiority of the FAC compared to the air force of its neighbour, and Colombia decided to purchase reinforcements.

Mirage 5COR FAC 3011 fresh from delivery in Colombia. (Dassault-Aviation, via Michel Liébert)

Mirage 5COA FAC 3022 taxiing at Bordeaux-Mérignac where the Dassault factory is located. (Dassault-Aviation, via Michel Liébert)

Colombia

Kfir C7/TC7 and Mirage M5COAM/CODM

The Kfir story

By mid-1970 the Rockwell engineers had become established in Israel and the idea was born to adapt the fuselage of IDF/AF Mirage IIICJ and Mirage 5 jets to accommodate the US-made General Electric J79-GE-17 engine. The same powerplant was used in the McDonnell Douglas F-4E Phantom II, which was in the process of entering IDF/AF service at the time. The US engine provided no less than 8,119kg (17,900lb) of thrust in full afterburner, compared to the 5,393kg (11,890lb) of the Atar 09C. In theory, the J79 would enable the Mirage to become a very potent fighter-bomber.

However, the idea proved much more complex to realise than expected, since the J79 engine not only had a greater diameter, but was also shorter and generated more heat than the Atar 09C. Adapting the Mirage airframe for the carriage of the engine initially proved beyond the abilities of IAI, even with the extensive support of the Rockwell Corporation. As a result, Gene Salvay requested help from Dassault, as well as from Lockheed engineer Ben Rich. Assistance was readily provided and the resultant aircraft – based on a Nesher airframe but named 'Ra'am B' – slowly came into being, recording its first flight on 4 June 1973. The Yom Kippur War of 1973 further postponed the start of series production. Eventually, the first of 27 examples of what would eventually became known as the IAI Kfir was completed in early 1975.

Although equipped only with the Elta EL-2001 ranging radar, the Kfir proved significantly heavier and less manoeuvrable than the Nesher, necessitating further aerodynamic refinement, foremost achieved through the introduction of small canard foreplanes. This resulted in the Kfir C1 variant, at least 25 of which were later leased to the US Navy and Marine Corps for use as adversary aircraft under the local designation F-21A Lion.

Series production of the Kfir C2 was launched in 1976, this variant including a redesigned wing, enlarged canards and much-improved avionics. No less than 185 aircraft were built, including a significant batch of Kfir TC2 two-seat conversion trainers. The TC2 retained full operational capability but carried less internal fuel. First flown in prototype form in 1982, TC2 production relied on fuselages most of which were again acquired from France. Subsequent assembly of the Kfir C7 variant brought the total number of airframes manufactured to more than 210, including the Kfir C10 variant that included additional hardpoints and some of the avionics developed for the stillborn IAI Lavi fighter project.

Although widely advertised, the Kfir initially achieved only limited export success, mainly due to the US refusal to grant permission to export J79 engines to specific countries. Ecuador purchased 12 newly built Kfir C2/TC2s in 1988, followed by four second-hand aircraft in 1996 and two newly built C10s in 1996. Colombia acquired 14 Kfir C2/TC2s (including a pair of airframes originally completed as C1s) in 1989. In addition, a significant number of Kfirs (without their J79 engines) were sold to South Africa in the late 1980s, for conversion to Cheetah C standard. In the 1990s Sri Lanka joined the Kfir club when it purchased a number of Kfir C2/TC2s, followed in 2000-01 by Kfir C7s.[1]

---

[1] Primarily based on the article 'The Designer of the B-1 Bomber's Airframe', by Joe Mizrahi, published in *Wings* magazine, Volume 30, No. 4, August 2000

Latin American Mirages

By 1981 negotiations had taken place between Colombia and IAI concerning the purchase of a batch of 12 IAI Kfir C2 fighters, with deliveries from 1982, and for the modernisation of existing Mirage 5s. In the event, the US vetoed the sale of the related engines, and the Kfir deal was cancelled. The embargo was lifted in October 1987, and in April 1988 the Colombian congress approved the deal. On 6 October a contract for 12 IAI Kfir C2s and 1 Kfir TC7 (a TC2 upgraded to TC7 standard) was signed. All the aircraft belonged to the Heyl Ha'Avir and were purchased for a total of 200 million US Dollars, to be paid partially with coal, as part of the Shibolet programme. An option for another 12 aircraft was not taken up. All single-seaters were to be upgraded to C7 standard and the plan also included the upgrade of all surviving FAC Mirage 5s to the same standard. Initially it was planned to retrofit the Mirage 5s with the General Electric F404 or the Kfir's J79 engine, but this scheme was abandoned due to the high costs involved.

Modernisation work began in 1988 at Arsenal Madrid in Bogotá. Upgrade began with the pair of two-seat aircraft, which received new avionics, canard foreplanes, refuelling probe and two additional pylons, among other changes. Single-seat Mirage 5 FAC 3029 was sent to Israel where new systems would be installed. The aircraft then underwent flight tests in Israel, including trials with the refuelling probe, during which it received fuel from a Heyl Ha'Avir A-4E Skyhawk. On 20 April 1989 the aircraft returned to its unit in Colombia.

The first Mirage 5COA to be upgraded in Colombia (with the new designation Mirage 5COAM) was FAC 3026. This was re-delivered in 1990 after a first flight performed in January 1989 had revealed some avionics problems caused by the generator. All single-seat Mirage 5 aircraft were upgraded at the FAC's maintenance facilities at Base Aérea Madrid, near Bogotá. The aircraft received a Kfir-style nose, with an Elta EL/M-2001 fire control radar, a Weapons Delivery and Navigation System (WDNS), HOTAS controls, canards (75 per cent the size of those on the Kfir), four new hardpoints (taking the total from five to nine), an aerial refuelling probe and a single high-pressure fuel loading point. Twelve single-seaters were upgraded, including the Mirage 5COR FAC 3011,

A Colombian Kfir of the first batch, still at IAI facilities before delivery.
(IAI)

FAC 3003 was Israel's first two-seat Kfir.
(FAC)

which lost its reconnaissance capability and was re-serialled FAC 3035. The other Mirage 5COR was lost in an accident, as were the Mirage 5COAs FAC 3023 and FAC 3032.

Since the FAC had no tankers in service, Boeing 707 FAC 1201 was modified to tanker configuration with Israel equipment, in order to refuel the upgraded Mirages.

Meanwhile, delivery of the Kfirs began on 28 April 1989, with the arrival of FAC 3045 (serial numbers for the new single-seat aircraft were FAC 3040 to FAC 3051). The single TC7 arrived in 1990 and received the serial number FAC 3003. Training of pilots and mechanics was conducted with the support of the Ecuadorian Air Force. The Kfirs entered service with Escuadrón 213, which shared its base with the Mirage fleet, and immediately became involved in operations against guerrillas and drug traffickers.

On 10 December 1990, Kfirs and Mirage 5COAM jets bombed the town of Casa Verde during an Operación Militar de Ablandamiento (Softening Military Operation). The aim of this offensive was to allow ground forces into the area as part of Operación Colombia, the objective of which was to destroy the myth of FARC supremacy in the area. The aircraft were organised into Cobra flight (Lieutenant Colonel Gonzalo Morales Forero, Major Flavio E. Ulloa E. and Captain Rafael Velosa A.), Apollo flight (Major Miguel Camacho M., Captains Eliades Moreno and Miguel A. Barrera D.) and Gorilla element (Captains Hugo Acosta and Juan C. Vélez). By the end of the operation they had undertaken 30 sorties and dropped 186 bombs during 15.25 flying hours.

First aerial refuelling

On 13 November 1991, the FAC performed its first air-to-air refuelling using its own aircraft and crews. The tanker FAC 1201 was crewed by Lieutenant Colonel Fernando Soler and the first aircraft to refuel was Mirage 5COAM FAC 3024, flown by Captain Miguel A. Barrera D. The refuelling was made at a height of 12,000ft (3658m) and a speed of 270kt. Subsequently, Major Juan C. Ramírez M., Captain Jorge A. Suárez, Majors Fernando Medrano J. and Flavio E. Ulloa E., Capitán Juan C. Vélez and Israeli Air Force Captain Ram Brier also refuelled.

Anticipating future internal conflict in the country, CACOM-1 ordered the first night bombing training sorties in 1992. Later, in 1994, Captain Juan Fernando Correa Hernández recorded a FAC speed record, flying at Mach 1.98 in Kfir FAC 3049.

A Kfir refuels from a Boeing 707. (FAC)

A Mirage 5COAM and a Kfir refuel from a Boeing 707. (FAC)

Combat operations continue

During the 1990s, guerrillas shot down one of the Mirage 5COAMs while it was flying a close support mission. Generally these kinds of missions were flown in support of Army, Police and Marines ground forces that were engaged in fighting the guerrillas. The nature of the conflict was such that on many occasions the operations were conducted very close to inhabited zones. This made air operations extremely difficult if the FAC was to achieve accuracy and reduce collateral damage. In light of this, and in order to increase the efficiency of the missions, a new upgrade programme was launched in February 2001. The latest modernisation included the installation of night vision systems, while the two-seaters received a Cockpit Laser Designator System (CLDS) to direct the new IAI Griffin laser-guided bombs, purchased in 2001. The FAC initially acquired 12 Griffin kits to install on its Mk 81, Mk 82 and Mk 83 bombs and these were followed by additional orders. The Griffin bombs were the first of their kind acquired by a Latin American country for large-scale operational use, and are used by the single-seater Kfirs and Mirages. In addition, Chilean-made Cardoen CB250K cluster bombs had been purchased by the end of the 1990s. The cluster bombs were used until 2009, when they were destroyed after Colombia signed the Olso Treaty preventing the use of this type of weapon.

Colombia

Two Mirages (nearest camera) and two Kfirs over Colombia. (FAC)

Shortly after the first Griffin kits arrived, tests were performed at the Chelenchele range using an inert bomb with the laser guidance kit installed. The tests were a complete success. On 31 October 2001, Lieutenant Joana Ximena Herrera became the first woman to fly the FAC Mirage 5. Occupying the rear seat of a two-seater, she was a navigator and weapons specialist during a reconnaissance flight over the Palanquero area.

In February 2002 the government of President Andrés Pastrana ordered a major offensive against the guerrillas in the Demilitarised Zone in the south of the country. The area had been in the hands of the guerrillas since peace talks began in November 1998. The talks had broken down by the end of 2001 and the government began to plan to occupy the zone through a large-scale offensive. Plans were accelerated after the hijacking of an aircraft on 20 February 2002, and 31 FAC aircraft, including Mirages, Kfirs, A-37s, Douglas AC-47s, North American Rockwell OV-10s and Tucanos, launched a major bombing offensive over different FARC targets. Particularly heavily hit were targets around the city of San Vicente del Caguán.

These operations saw the first operational use of the FAC's Griffin bomb. After careful mission planning, a Mirage 5COD illuminated the selected target, a bridge over the Cafre river, which constituted one of the main access routes between the Demili-

Kfir TC7 FAC 3003 displays the new grey paint scheme. (Simón Cuartas)

143

Latin American Mirages

A Mirage 5COAM flies alongside the Boeing 707 tanker.
(FAC)

A Kfir over Colombia.
(FAC)

Two Mirage 5COAMs in flight.
(FAC)

Colombia

tarised Zone and the rest of the country. The guerrillas had used the bridge to launch terrorist attacks before quickly finding refuge inside FARC-controlled territory. A Mirage 5COAM launched the bomb successfully, destroying the bridge. Subsequently, the laser-guided bombs began to be used regularly in attacks against the guerrillas. Both Mirages and Kfirs waged a constant bombing campaign against bridges, drug-production laboratories, runways, camps and other facilities.

As combat operations continued, in early 2003 the Spanish Air Force offered to donate eight ex-Qatari Mirage F.1s to reinforce the FAC's combat capacity. After a major discussion involving the defence ministry and the FAC, the Spanish government decided to withdraw the offer. In the event, the FAC operations launched in 2002 led to a reduction in guerrilla activity, which led to a consequent reduction in the tempo of air strikes.

A Kfir over the mountains, equipped with Python III missile training rounds.
(FAC)

Kfir down

On 4 June 2003 the FAC suffered the loss of a second Kfir. FAC 3046 crashed into the Magdalena River, 16 miles (26km) from its base, after a birdstrike damaged the engine. Captain Juan Manuel Grisales recalled that he was flying at 5,000ft (1,524m) and at a speed of around 700km/h (435mph) when he felt a powerful impact. The pilot checked the instruments and noted that the aircraft was losing power. He called his wingman and informed *'Apache, Apache! I'm losing power, a bird is in the engine'*. His wingman guided him through the correct procedure to restart the engine, but the engine refused to respond. The next order was to eject the external fuel tanks to reduce weight and drag, but the aircraft was still losing height. When the pilot realised he was now at a very low altitude, he made the decision to eject.

The pilot recovered in the river, close to his aircraft. Since the current was very powerful, and was carrying him away, he released his parachute and started swimming

Grey-painted Kfirs in flight over Colombia.
(FAC)

Overall-grey Mirage 5COAM FAC 3030 with the Grupo de Combate 11 logo on the nose and colourful national roundel. (Author's archive)

to the shore. However, the weight of his equipment made the task almost impossible and the parachute was entangled with his legs. In a bid to survive, he reached the seat cushion and clung to it. Some 10 minutes later he saw a motor boat approaching – four fishermen from a local town had seen the aircraft go down and went to the pilot's rescue as fast as they could.

The larger of the fishermen took the pilot by his arms and out of the water, with the parachute and equipment still with him. The pilot was taken to the town of Buena Vista, where called his wingman and let him know he was alive and safe. Immediately, a helicopter took off from Palanquero with a search and rescue team, while Captain Grisales went with the people of the town to look for a place where the helicopter could land, finding a football field. With the help of locals, the pilot was finally picked up by the helicopter and taken to the base.

Kfir C7 FAC 3040 armed with two Griffin LGBs. (FAC)

Colombia

Overall-grey Kfir FAC 3044 displays several colourful badges: the squadron badge is on the tail, with Grupo de Combate 11 and Kfir logos on the nose.
(Author's archive)

War continues

Since 2002 the guerrillas have never again taken control of large areas of territory. However, guerrilla attacks remained constant, and the FAC continued its bombing and close air support missions using all available resources. Between 2003 and 2007 the Mirage 5 and Kfir completed an average of three weekly missions per squadron, attacking the guerrilla forces attempting to occupy towns. To avoid collateral damage, pilots are required to be completely certain they have positively located the target before firing. This increases the risk of the missions, especially since the aircraft are usually received by intense anti-aircraft fire. Furthermore, operations are flown among complicated Colombian geography, with mountains covered by dense jungle, and since 1997 most missions have been flown at night. *'From the ground they call us with the radio and in the middle of the fight we try to put our weapons where they need them'*, said Major Guillermo Olaya, one of the 11 Kfir pilots equipping Escuadrón 212 by 2004.

Seen here in Israel, FAC 3004 was one of the new two-seaters sold to Colombia. Prior to delivery, the aircraft crashed during a test flight while attempting to take off from Cartagena on 20 July 2009. IAI agreed to replace the aircraft.
(IAI)

147

Latin American Mirages

FAC 3055 was the first modified Kfir C10, and is seen here with a Litening pod and the new radar-nose.
(IAI)

On 20 July 2004 the first Rafael Python III air-to-air missile was launched from a FAC Kfir against a target over the Colombian zone of the Pacific Ocean. The test launch was a complete success. The new missiles had been received shortly before in order to increase the combat potential of the Kfir and Mirage 5.

By the beginning of 2008 the FAC had announced the signing of a contract for the purchase of 11 Kfir C7 and 2 TC7 aircraft. Ultimately the plan was to use these new jets to replace the Mirage 5. The contract included the upgrade of all single-seat Kfirs, including the 11 still in FAC service, to C10 standard and C12 standard. The two-seaters (TC 12 standard) were to receive similar modifications, albeit without the radar.

The first example, FAC 3054, was handed over in Israel in April 2009, before having undergone modernisation. Soon after, deliveries began in Colombia. One of the new two-seaters, FAC 3004, was lost on 20 July 2009 when taking off from Palanquero for a pre-delivery test flight with Israeli pilots. When the engine failed during take-off, the aircraft ended up crashing into rocks very close to the sea. The pilots escaped the aircraft with minor injuries, but the Kfir was lost. IAI agreed to replace the aircraft since it had been lost prior to delivery. In September 2009, flight tests of the first modified Kfir C10 began. The aircraft, FAC 3055, was fitted with a Litening pod for target designation.

Newly delivered Kfir C10 FAC 3059 during the 90th anniversary celebrations of the FAC.
(Javier Franco 'Topper')

148

Colombia

Map of Colombia

Chapter 5

ECUADOR

Mirage F.1JE/BE

The Fuerza Aérea Ecuatoriana (FAE, Ecuadorian Air Force) was one of the best-equipped Latin American air forces in the years after World War II. It was the second air arm in the region to receive English Electric Canberra bombers and the third to operate Gloster Meteor fighters. In 1954, Ecuadorian air power was comparable to that of Peru, Ecuador's eternal adversary, which had occupied part of Ecuador's Amazonian territory during the Ecuadorian-Peruvian War of 1941.

The FAE's Meteors were reinforced by the Lockheed F-80C Shooting Star in 1958, which formed Grupo de Combate 211, part of Ala de Combate 21 located at Base Aérea de Taura, close to the city of Guayaquil.

By the beginning of the 1970s the FAE was planning a programme of modernisation in order to balance the power of the Peruvian Air Force, equipped with the modern Mirage 5P. A first step was the purchase of the BAC Strikemaster for advanced training and light attack, and the SEPECAT Jaguar for ground attack. In so doing, Ecuador became the sole American operator of both models. The British-supplied equipment was complemented by 12 A-37B Dragonfly ground-attack jets, which initially equipped Grupo de Caza 212. Despite these acquisitions, Ecuador still lacked a modern fighter, and in 1975 discussions began with IAI with a view to buying 24 Kfir C2s. A contract was signed in 1976, but the US government of Jimmy Carter vetoed the sale of the Kfir's J79 engine. Instead, the Israelis offered a batch of second-hand Mirage IIIC fighters, but this offer was rejected by the FAE.

Denied the Kfir, Ecuador turned to Europe to look for a replacement for the Meteor and T-33. The initial plan was to purchase a fighter from the Mirage III/5/50 family, but Ecuador finally settled on a batch of 16 single-seat Dassault Mirage F.1JA fighters (designated F.1JE for Ecuador) and a pair of two-seat Mirage F.1BE combat trainers. The Mirages would be equipped with the Cyrano IVM radar, providing the capability to detect very low flying threats.

The purchase contract was signed in 1978 and on 1 August the first FAE pilots (Majors Héctor Heredia and Patricio González; Captains Luis López, Gustavo Bucheli and Marco Estrella and Lieutenant Hernán Ayala) and mechanics went to Reims, France, for a five-month theoretical course. On 29 January 1979 the FAE pilots were posted to Orange, where they began flight-training missions on the Mirage F.1, which included a total of 20 missions and 294 flying hours. On 13 February Dassault delivered the first four Mirage F.1s (serial numbers FAE 801 to FAE 804) to the FAE. On the same day, using the Cóndor callsign, the single-seaters FAE 802, FAE 803 and FAE 804 took off with French pilots from Dassault facilities to Orange, where they were used to train the Ecuadorian pilots.

After two months of activities, each Ecuadorian pilot had completed 20 flying hours on the Mirage F.1 in France and on 27 April 1979 they returned to Base Aérea de Taura where they would begin a new era of operations with Escuadrón 2112. The aircrew had to wait for their new equipment to arrive, however, with the Mirages being disassembled and sent by ship to Ecuador.

By May 1979 the first four aircraft had arrived in Guayaquil harbour and assembly began. The first flight of a Mirage F.1 (FAE 804) in Ecuador took place on 26 June, with French pilot Jean Bongiraud at the controls. Two days later Grupo de Caza 212 was declared operational at Taura with Major Hernandez as commander. Trained in France, Captain Gustavo Bucheli was the unit's first instructor, and he was followed by Captains Marcos Estrella and Hernán Ayala.

The Mirage F.1JEs carried the serial numbers FAE 801 to FAE 831, while the Mirage F.1BEs were FAE 830 and FAE 831. The remaining aircraft were delivered during the course of 1979, being completed by the arrival of the two-seaters in 1980. Also delivered were a batch of Matra R.530 and R.550 Magic air-to-air missiles. On 25 June 1980, FAE 804 was lost in an accident. During 1980, Ecuadorian President Jaime Roldós Aguilera flew in one of the two-seaters, becoming the first president to do so. Unfortunately, he was subsequently killed in an air crash on 24 May 1981.

Tensions with Peru

On 22 January 1981, following reconnaissance sorties flown by Peruvian helicopters over the Condor Cordillera, which divides Peru and Ecuador, conflict between the two countries broke out. Following the reconnaissance missions, Peru reported that some of its border surveillance posts that had been abandoned were now occupied by Ecuadorian troops, and that the latter had raised their country's flag.

Mirage F.1JE FAE 812 armed with two Python III and two R.550 Magic missiles. The Python III has an all-aspect capability, longer range, and a heavier warhead than the R.550 Magic.
(FAE Vía Roberto Bertazzo)

A Mirage F.1 and two Kfir CEs over Guayaquil. (FAE)

Peru immediately began to deploy forces to the region and performed attack missions in the following days. In particular, use was made of the A-37B. According to the Dirección de Inteligencia of the Peruvian Air Force, the Ecuadorian order of battle included 12 Jaguars (of which 6 were operational), 17 Mirage F.1s (11 operational), 5 T-33s (3 operational), 12 A-37Bs (10 operational), 11 Strikemasters (8 operational) and 3 Canberras (only 1 operational). Grupo de Caza 212 was divided in two, with one component remaining at Taura and the other deployed to Manta air base. Both components kept pilots on five-minute alert.

During a patrol tasked with the escort of a Bell 212 helicopter on a support mission, the Mirages of Apache flight, comprising Major William Birkett and Captain Eduardo Carrera, were attacked by Peruvian SAMs, but the pilots managed to evade the missiles.

Mirage 5P fighters covered numerous Peruvian attack sorties, but in response the FAE only used its A-37Bs to bomb Peruvian positions. By the end of February the conflict had ended, following mediation by other countries in the region. The FAE Mirage F.1s were kept on alert but saw no combat. Only one Jaguar made a successful electronic intelligence mission to verify the position and range of the Peruvian early warning radars.

When Ecuador purchased Kfirs in 1982, Grupo 212 was divided into three squadrons: Escuadrón 2111 with the Jaguar, Escuadrón 2112 with Mirage F.1s and Escuadrón 2113 with Kfirs.

Meanwhile, on 7 March 1983, another accident occurred when Captain Hernán Ayala suffered an engine fire while flying Mirage F.1 FAE 810. The pilot tried to return to base to save the aircraft, but as he was about to land he lost control and was forced to eject. This was the first ejection in the country from a supersonic aircraft, and it saved Ayala's live.

By 1984 the Mirages had received chaff and flare dispensers in order to increase their survivability.

On 18 June 1985 the Centro de Operaciones Sectoriales No. 1 (1st Sector Operations Centre) detected an aircraft flying in prohibited airspace. Piraña flight was scrambled at 11.15, with the Mirages piloted by Captains Wilson Salgado and Gustavo Cuesta. The flight found a twin-engined Beechcraft and attempted to contact the crew. The Mirage pilots signalled to the crew to show they had been intercepted, but the Beechcraft responded by reducing speed and attempting to escape in the clouds. When it became clear that the Beechcraft intended to escape, the interception control officer ordered the Mirage pilots to open fire. The lead Mirage shot down the Beechcraft using the onboard cannon. This was the first aircraft shot down by FAE Mirage F.1s.

Mirage F.1JE FAE 815 was lost in May 1985, followed on 21 January 1988 by Mirage F.1JB FAE 832.

In the following years there were regular encounters with Peruvian aircraft over the border zone. FAE aircraft involved included both the Mirage F.1 and Kfir, and it was an example of the former that launched a Magic missile against a Peruvian aircraft, albeit unsuccessfully. The Mirages received RWRs in addition to their chaff and flare dispensers and were prepared to deliver Israeli-built bombs.

By 1994 the fleet was significantly reduced, with only 13 Mirage F.1JEs and 1 Mirage F.1JB in service, complementing 7 Kfir C2s and 2 Kfir TC2s. By that time, both types of fighter had begun to be used to intercept illegal flights, usually carrying drugs. In such missions, the FAE fighters were cleared to shoot down aircraft declared to be undertaking illegal activities. During one of these interceptions, two Mirage F.1JEs intercepted a twin-engined aircraft and asked the pilot to follow them and land. When the civilian pilot refused these orders, the aircraft was shot down by the lead Mirage F.1 using the onboard cannon.

Cenepa War

The continuous border incidents concerning the Condor Cordillera led to a new spate of tensions around the Cenepa River by the mid-1990s. On 12 December 1994, after claims that Ecuadorian troops had again occupied their border posts, Peruvian troops requested the Ecuadorian soldiers to evacuate the posts. The Ecuadorians denied the Peruvian request to leave and on 11 January 1995 clashes took place between both armies. On 24 January, Ecuadorian aircraft conducting a reconnaissance mission discovered a helipad on the northern end of the Cenepa River. This was later named

Base Norte and according to the Ecuadorians, the base was within their territory. The Ecuadorians ordered the eviction of the Peruvian forces. The FAE immediately organised a defensive disposition under the command of the Comando Aéreo de Combate (COMAC). This included the formation of the Grupo Aéreo Amazonas, with transport and combat units of Ala 22 and 23, which were deployed to the conflict zone for close air support, transport and rescue. The disposition also included collaboration with Army and Naval air assets, in order to fulfil maritime surveillance and other missions. Finally, Ala de Combate 21 at Taura received the order to provide air cover with its Mirage F.1 and Kfir C2 fighters, while the Jaguars were prepared for a strategic air strike against Peruvian targets, in case the conflict developed into open war.

On 27 January, the Peruvians launched more significant attacks and on the following day the Ecuadorians destroyed Base Norte, forcing the Peruvians to retreat. On 29 January the Peruvians launched a major attack against all Ecuadorian border posts along the Condor Cordillera, and Peru lost two Mil Mi-8 helicopters. Ecuador tried to maintain a defensive posture, in order to highlight Peru as the aggressor. Both countries kept the conflict within the Cenepa River zone and open war was avoided.

On 31 January the nations brokering a peace deal between the two countries (Argentina, Brazil, Chile and the US) announced a ceasefire. Ecuador declared that it agreed to the terms of the ceasefire, but on the following day fighting resumed.

On the evening of 6 February the Mirage F.1 and Kfir had their first contact with Peruvian aircraft. However, contact was made from a considerable distance and the FAE jets were unable to engage the Peruvian aircraft. The latter had completed an attack and had just crossed the border when they were detected. The FAE had been ordered not to cross the Peruvian border, and the intention was only to defend against Peruvian attacks. The situation was repeated on the following day, when the Kfir and Mirage provided cover for the first Ecuadorian air strikes against Peruvian positions, these being carried out by the A-37B.

By 9 February, the Fuerza Aérea Peruana (FAP) intensified its actions over the Cenepa with 16 air strikes over Ecuadorian positions, using Mirage 5P and Su-22M jets during the daytime, and Canberra B.Mk 68 bombers by night. Follow-up raids were flown the next morning by A-37Bs and Su-22Ms.

Mirage F.1JE FAE 807 carrying French-made Magic and Israeli Python III air-to-air missiles. (Rogier Westerhuis)

Latin American Mirages

Ecuadorian victories

On 10 February the FAP continued its air strikes against Ecuadorian positions, with A-37Bs and Su-22Ms flying from Base Aérea de Piura. Up until now, the FAP was yet to encounter any opposition from the FAE. The reasons for this were the bad weather over the war zone and that fact that, by the time FAE fighters arrived on scene, the Peruvian aircraft had already crossed the border.

At 12.42 on 10 February, 'Halcón' station of the Ecuadorian air defence system informed COMAC that five targets had been detected, heading towards the conflict area. Two of the targets were flying at a speed of 400km/h (249mph) and three at 300km/h (186mph). At Taura, the order was given to scramble at 12.49. The order to scramble was directed to Conejos flight of Escuadrón 2112, consisting of Major Raúl Banderas in Mirage F.1 FAE 807 and Captain Carlos Uzcategui in FAE 806. The Mirages attained an altitude of 30,000ft (9,144m), flying at a bearing of 175 degrees.

At 12.53 the operations centre in the town of Patuca received information concerning a possible air strike against them in the next 12 minutes. Meanwhile, two A-37Bs were put on two-minute alert at the airport. Two minutes later, COMAC notified an FAE Beechcraft T-34C Turbo Mentor, acting as a forward air controller over the town of Méndez, that it was to order all other Ecuadorian aircraft to leave the theatre of operations. This would leave the Mirages and Kfirs with the freedom to fire against any

Captain Carlos Uzcategui with Mirage F.1 FAE 806 that used to shoot down one of the Peruvian Su-22Ms.
(FAE)

156

aircraft they found. On arriving in the area, the Mirage F.1s were ordered to descend to 20,000ft (6,096m) and look for bandits.

Four minutes later the Mirage F.1s detected two Su-22Ms, first with their radars, and later visually. The FAP jets were flown by Comander Víctor Maldonado and Major Enrique Caballero Orrego. Using the callsign Poeta, then Major (now Colonel) Raúl Banderas was flying at 2,000ft (610m), some 15 miles (24km) from the Sukhois. He remembers what happened. *'They gave us the heading, distance and height of the targets. We took that course and eight minutes later I had a radar contact in the direction where they were expected to be. I ordered my wingman to change to attack formation and we continued until we could visually identify them, with the intention of determining how to attack them. Anyway, we decided to approach closer to be sure there were no other enemy aircraft in the area, in order to prevent being attacked.'*

The Su-22s were heading to the Ecuadorian border post at Tiwintsa, in the middle of the jungle, and the Peruvian jets descended to 1,800ft (549m). *'In my headphones I heard the sound indicating my Magic missile was ready to fire'*, recalls Banderas. *'We put ourselves at their 6 o'clock. When they saw us and started to manoeuvre it was too late, because I had fired my first missile. The missile hit one of the Sukhois, the wingman (commanded by Major Caballero), and immediately I gave the order to my wingman to attack the other aircraft. He hit the target. The Sukhois are very strong aircraft. While all this was happening, I heard all the time that my RWR was indicating we had a threat. When we were attacking, other aircraft appeared behind us – we didn't see them, but we detected their electronic threat. This forced us to hurry to accomplish our mission. In order to finish the action I fired a second missile that hit the enemy aircraft in its mid-section. It exploded in a great ball of fire and then fell into a spin, and the pilot ejected. My wingman also fired again and hit the other aircraft. We left almost at supersonic speed and very low over the trees to protect ourselves from the threat we had behind us, and we managed to return to our*

The badge on FAE 807 to commemorate its Sukhoi kill. (Roberto Bertazzo)

FAE 806, flown by Captain Carlos Uzcategui, is one of the jets that shot down a Peruvian Su-22M on 10 February 1995. (FAE)

base. The threat was present for about 30 seconds, so we dropped chaff and we lost them. We made our return safely'. Both Peruvian pilots, although they ejected, died shortly after touching down. The Peruvian jets were shot down at 13.05. The signal on the RWR of the Mirage F.1s came from two FAP Mirage 2000Ps, but these failed to impede the loss of both Su-22s. Both Mirage F.1s returned to their base, after becoming the first Ecuadorian aircraft to achieve a victory in air-to-air combat.

Thereafter the FAP presence over the conflict zone diminished, but fighting on the ground continued and the FAE continued to launch air strikes, with the A-37B covered by the Kfir and Mirage F.1.

On 13 February the Peruvians launched a land offensive, but they could not defeat the Ecuadorian defences, and no important air activity was recorded. On the following day both countries agreed to a ceasefire. However, from 14-21 February some small clashes took place, these involving patrols from both countries. The order was given to both nations to return to their original positions and a demilitarised zone was declared in the former conflict zone. A final combat took place on February 22, leaving many dead on both sides, but no subsequent major clashes occurred, and with that the conflict ended.

Return to peace

After the war, Rafael Python III air-to-air missiles were purchased and the Mirage F.1s were modified to carry these weapons on their underwing pylons. The Pythons could not be carried on the Mirage F.1's wingtip hardpoints, so the Magic continued to be used on these stations.

In late 1997 an incident occurred when two Mirage F.1s intercepted a US Navy P-3 Orion that was flying along the Ecuadorian coast on a counter-narcotics mission. The Ecuadorian pilots tried unsuccessfully to communicate with the crew of the P-3, which was flying within Ecuadorian airspace. As a result, and following international inter-

FAE 830, the last remaining Mirage F.1BE two-seater. (Stefano Rota)

A Mirage F.1 with Durandal bombs, Python III and R.550 Magic missiles.
(Roberto Bertazzo)

ception procedure, the Mirages fired on the intruder, which was damaged and immediately landed at an Ecuadorian air base. According to the US Navy, the Orion was flying 61km (34 miles) from the coast, while the FAE reported that the aircraft was only 9km (6 miles) from the coast.

Since then the Mirage fleet has experienced no further accidents or incidents, but budget cuts grounded most of the aircraft in the new millennium. The Mirage F.1 will also be partially replaced by six Mirage 50EV/DV fighters donated by Venezuela and delivered in late 2009 and early 2010. In 2009 the Mirage F.1 marked its 30th anniversary in Ecuadorian service and FAE 806 was painted in a special scheme to commemorate the fact. Currently there are no plans for urgently required upgrades for the Mirage F.1s, however, some spares are being received from Venezuela, since certain Mirage 50 components are compatible with the Mirage F.1.

Mirage F.1 FAE 806 received a special paint scheme in 2009 to commemorate the 30th anniversary of the type in Ecuador.
(Author's archive)

Kfir C2/TC2

Based on the lessons of the 1981 conflict with Peru, the FAE determined it needed to increase its power. Once again, Ecuador examined the purchase of a batch of IAI Kfir C2s. By this time the US government had lifted its embargo and Ecuador was able to sign a contract on 21 May 1981 for 10 single-seaters (serial numbers FAE 901 to FAE 910) and a pair of two-seat Kfir TC2s (FAE 930 and FAE 931). These were the first export examples of the Kfir and were the only aircraft sold new (Colombia and Sri Lanka, the other export customers, acquired them second hand). In common with the Israeli Kfirs, they were equipped with the Elta EL/M-2001B radar and an Israel Electro-Optics panel for data presentation, together with a digital inertial navigation system.

The contract also included support equipment, spares, training, weapons (including Shafrir II air-to-air missiles) and a flight simulator. On 16 July 1981 the first group of pilots went to Israel to receive training on the new model, which began on 11 August. Further pilots and technicians later followed this initial cadre.

When the first group of pilots returned to Ecuador, on 10 March 1982, the commander of the FAE reactivated Escuadrón 2113, which had previously flown the Strikemaster. From 25 March the new jets began to arrive in Ecuador. FAE 905 made the first flight of a Kfir in Ecuador on 31 March, with an IAI pilot at the controls. On 19 April 1982 a first flight by an Ecuadorian pilot took place, with Major Hernán Quiroz in the cockpit; this date is remembered as the anniversary of the squadron. On 11 June 1982, the president of Ecuador, Oswaldo Hurtado Larrea, declared the unit operational.

In 1985 the excellent level of training attained by the crews of Escuadrón 2113 was demonstrated when they won the Copa Taura gunnery competition, against the Jaguar and Mirage F.1, an achievement that would be repeated in following years.

Air combat

Border incidents with Peru were by no means unusual, but one which occurred in 1985 was the first in which the Kfirs participated. On this occasion, two Su-22s of Escuadrón de Caza 11 'Los Tigres' of the Fuerza Aérea del Perú (FAP, Peruvian Air Force) crossed the border and began flying within Ecuadorian airspace at low level and in combat for-

Kfir FAE 907 in flight. (FAE)

Kfir FAE 907 is serviced in a HAS.
(Author's archive)

mation. They turned and started flying along the border, but remained some 16km (10 miles) within Ecuadorian territory. The Peruvian pilots were alerted by their RWR that an Ecuadorian surveillance radar had detected their aircraft. In response, the FAP aircraft descended further. The Ecuadorian radar operators managed to track them intermittently, and a pair of Kfirs flying on patrol close to the area was sent to investigate.

The Kfir pilots detected the Sukhois at a distance of 8km (5 miles) and began to descend fast behind the intruders. At maximum missile range and with an angle of 90° to the target, the lead Kfir pilot launched a Shafrir. This missed, and the missile exploded harmlessly after reaching its maximum range.

The Sukhoi pilots, now alerted to the presence of the Ecuadorian Kfirs, turned hard to the right and escaped with afterburners at maximum speed. Both Su-22s had left Ecuadorian territory in a matter of seconds. The Kfirs returned to base after becoming the first FAE aircraft to launch an air-to-air missile in a combat situation. In the following years, such encounters were regular, and also included the FAE's Mirage F.1 fighters.

A FAE Jaguar is escorted by a Kfir and a Mirage F.1, illustrating Ecuador's three main combat types during the 1980s and 1990s.
(FAE)

A Jaguar, a Mirage F.1 and a Kfir fly alongside an A-7E Corsair II during an exercise with the US Navy.
(FAE)

The first Kfir loss also occurred in 1985, when on 7 February FAE 910 suffered a birdstrike that hit the windshield. The pilot, Lieutenant Marco López, ejected safely over Lomas del Sargentillo.

Accidents continued, with another Kfir being lost on 24 February 1988. On 3 May 1989, Kfir FAE 904 was lost after an engine fire during take-off. The pilot requested deployment of the barrier at the end of the runway, but this was not erected in time. The jet continued off the end of the runway and into trees. The pilot was killed and the aircraft was completely destroyed.

By the end of the 1980s the Kfirs had received chaff and flare dispensers and RWR.

Another accident occurred on 12 August 1994 when Kfir C2 FAE 903 suffered an engine fire over Santa Helena peninsula, with Major Juan Vivero at the controls. The pilot ejected safely. The fleet was by now considerably reduced, with only seven Kfir C2s and two TC2s. These were operated alongside 13 Mirage F.1JEs and 1 Mirage F.1JB. By that time, both fighters had begun to be used to intercept illegal flights, usually aircraft carrying drugs. Rules of engagement meant they were cleared to shoot down aircraft conducting illegal activities.

Kfir at war

On the same day that FAE Mirage F.1s scored aerial victories against Peruvian Su-22s, the Kfirs were also successful. On 10 February, Peruvian A-37B Dragonflies and Su-22Ms were detected crossing the border. A pair of Mirage F.1s was immediately scrambled from Taura at 12.49. They were followed one minute later by Bronco flight from Escuadrón 2113, consisting of Captain Mauricio Mata in Kfir FAE 905 and Captain Guido Moya in FAE 909.

Ecuador

Two Kfirs and a Mirage F.1 (centre) during a formation break.
(FAE)

While the Mirage F.1s shot down the Peruvian Sukhois, the Kfirs remained on combat air patrol close to the conflict zone. Captain Mauricio Mata remembers the mission. *'Once we were over Numbatcaime on Alto Cenepa, we stayed for about 15 minutes. Our main mission was to make visual contact with the enemy. Over the radio we heard the combat between the Conejos (Mirage F.1) and the two Peruvian Su-22s, which were shot down close to Gualaquiza. With this in mind, and close to abandoning our CAP station, Captain Moya saw two aircraft and immediately informed me. With the targets in sight I started the pursuit, looking for a good position to fire and shoot them down. My wingman stayed behind to give me support. I confirmed the aircraft were enemy, on account of their behaviour: completing evasive manoeuvres and ejecting their loads to increase manoeuvrability.*

'Once the target (an A-37) was in sight I fired a missile (a Shafrir II). Fortunately the missile left my aircraft fast. It left a contrail that stopped when it had found the enemy aircraft, which broke in two with the impact. Once the missile hit, no ejection was seen. The aircraft descended out of control until I lost it in low clouds.'

The A-37 was piloted by Commander Hilario Valladares and Major Gregorio Mendiola. Meanwhile, the wingman, Commander Fernando Hoyos, made his escape by manoeuvring at very low level. Valladares remembers the mission. *'I had no mission assigned for that day. Because of my rank as a commander I requested to be assigned to a mission. A young pilot had to give up his place, since I wished to lead by example'*. Valladares' mission that day was to bomb a strategic target, a Peruvian border post at Tiwintsa that was occupied by Ecuadorian forces. *'We flew to the target. It was about 13.00. We were close when we saw two Ecuadorian Kfirs. Despite that, we decided to continue with our mission. We arrived at our target and dropped the bombs. The Kfirs detected us and they turned towards our aircraft. Everything happened in seconds. The Kfirs were pursuing us; I decided to try to evade them. I tried*

Captain Mata alongside the Kfir in which he shot down a Peruvian A-37B on 10 February 1995.
(Roberto Bertazzo)

The kill marking on Kfir FAE 905. (Roberto Bertazzo)

to reach the minimum height to get into the canyons and avoid being shot down. The Kfir was much faster and reached us. I couldn't escape and I had no choice but to turn and face him to prevent him from taking aim. Then I saw the missile. I manoeuvred but it hit me and damaged the tail, destroying the engines and the rudder. The aircraft entered an inverted corkscrew, out of control.

'I thought of my family and tried to control the aircraft, but I couldn't. I saw Major Mendiola leaving the aircraft – he had ejected. I was trying to save my aircraft – I couldn't lose it. I wanted to control it and take it to Piura again. I saw the

Kfirs FAE 906 and FAE 909 armed with 500lb (227kg) bombs.
(FAE)

164

ground moving closer and then I pulled the ejection handle. I couldn't see where my aircraft had fallen. It was the first time I had descended with a parachute. I had no fear, only the thought of escaping safely. My first intention was to go down in the river – I couldn't. The parachute took me to the shore, and bushes, rocks and trees. I hit a big rock with my back, but I was miraculously unharmed'. Meanwhile, the Kfirs returned to their base.

Upgrade and new acquisitions

Although the FAE did not lose any combat aircraft during the conflict with Peru, immediately after the conflict ended it was clear that Peru's air force possessed a qualitative superiority. In case of a large-scale war, Peru's Mirage 2000P, Mirage 5P, Su-22 and Canberra aircraft, among others, were likely to dominate the FAE. Also of concern was the fact that Peru immediately began negotiations after the war to buy a batch of MiG-29s and another of Su-25s to further increase its air power.

Based on these developments, the FAE entered negotiations with IAI to upgrade the Kfirs, acquire new weapons and increase the inventory. As a first step, Python III air-to-air missiles were acquired to replace the Shafrir on the Kfir. Furthermore, three second-hand Kfir C2s and a single Kfir TC2 were purchased, these receiving the serial numbers FAE 911, FAE 912, FAE 913 and FAE 932 respectively. The three single-seaters arrived from 1996 and featured some minor modifications, the most important being a one-piece windshield. Meanwhile, a more ambitious plan began, with the aim of upgrading the Kfirs to Kfir 2000 standard.

A Jaguar, a Mirage F.1 and a Kfir fly with an F-16 during an exercise with the USAF. (FAE)

Kfir FAE CE 915 armed with Python III missiles.
(Rogier Westerhuis)

Caza Ecuatoriano

After the cancellation of the IAI Lavi in 1987 and the lack of interest in the IAI Nammer (a Kfir with the General Electric F404 engine as used on the F/A-18), IAI elected to modernise the existing Kfir C7. The result was the Kfir C10 version, which included many components from the two previous projects. With customers showing interest in the Kfir C10, the company developed the Kfir 2000 project. This included some structural modifications, permitting a minimum of 5,000 flying hours, modifications to the engine, a new aerial refuelling probe and a completely new forward fuselage. The cockpit was totally redesigned, with two 5in multifunction displays, new HUD, HOTAS controls, a new electrical system with only one third of the weight of the former, and a helmet-mounted display (HMD) so the pilot could aim the weapons by only moving his head. However, the main change was the installation of an IAI Elta EL/M-2032 multimode radar, converting the Kfir into a true multi-role fighter, since it offered both air-to-air and air-to-surface modes. Also inherited from the Kfir C10 was a Mission Planning Centre (MPC), which allowed mission data to be prepared at the base on a PC before being transferred to the aircraft via a Data Transfer Cartridge (DTC). Planned weapons included Rafael Python IV and Derby air-to-air missiles, IAI Griffin laser-guided bombs combined with a Litening pod, as well as other options. Eventually, only the Python IV was acquired.

IAI modified two Kfirs to this new configuration, 871 and 896, but the second was used as a ground demonstrator for display at exhibitions, and was fitted with a transparent nose.

Following the sale of the four additional aircraft to Ecuador, in 1998 IAI and the FAE negotiated the upgrade of 8 of the 10 single-seaters to Kfir 2000 standard (known as Kfir Caza Ecuatoriano, or Kfir CE, meaning Ecuadorian Fighter). At the same time, IAI would sell the demonstrators to the FAE, since no other buyers could be found for this version.

By 1999 the two demonstrator aircraft had been sent to Ecuador. Serial number 871 become FAE 914 and serial number 896 become FAE 915. Meanwhile, at the FAE workshops in Lacatunga, upgrade work had begun on the Kfirs already in service with

Ecuador

Two Kfir CEs each armed with one Python III and one Python IV.
(Rogier Westerhuis)

the FAE, starting with FAE 901. This aircraft was officially delivered at the beginning of 2001 and with it the HMDs were also delivered, before the first Python IV missiles, purchased with the same contract and delivered in the same year.

Although the upgraded aircraft were fitted with refuelling probes, Ecuador has no tanker capability, despite having discussed the purchase of a tanker with Israel as part of the Kfir upgrade deal.

Ecuador subsequently upgraded FAE 902, FAE 906, FAE 908 and FAE 909. Serial number 913 was lost in an accident on 24 April 1998 when flown by Teniente Marco Palacios. The aircraft hit a bird and the pilot was forced to eject. The remaining aircraft (FAE 905, FAE 907, FAE 911 and FAE 912) were not modified as a result of budget cuts,

Parked under a sunshade, Kfir CE FAE 901 is in intercept configuration, carrying Python AAMs and a centreline fuel tank.
(Stefano Rota)

leaving seven Kfir CEs, four Kfir C2s and three Kfir TC2s in service. However, FAE 931 was lost in an accident on 21 October 2004. The accident occurred after take-off, when the pilot detected a problem and decided to return to base. While returning, the engine caught fire. Attempting to reach the runway, the pilot ejected the external tanks, one of them landing on a school – fortunately without causing any injury. As he was about to land, the pilot, Capitán Alez Padilla, lost control of the aircraft. Padilla, together with Captain Patricio Velazco in the rear seat, ejected safely.

A major reduction in the defence budget at the beginning of the new millennium affected the Kfir as well as the Mirage F.1. Most of the Kfir fleet was grounded, but by 2010, the new government of Rafael Correa was making efforts to restore the fleet to operational capability.

Two Kfirs accompany a Mirage F.1.
(Rogier Westerhuis)

Mirage 50EV/DV

From the beginning of the new millennium, budget cuts severely affected the FAE. By 2009 almost all FAE Mirage F.1 fighters had been grounded, with no more than four in service, and these only flying sporadically. In order to return some of these aircraft to service, the FAE asked the Venezuelan Air Force for spares for the engines – Venezuela operated the Mirage 50, which uses the same engine as the Mirage F.1. In return, the Venezuelan government offered to donate six airworthy Mirage 50s as well as to provide spares for the Mirage F.1s.

Mirage 50DV 7512 taxies for take-off for its delivery flight to Ecuador on 26 October 2009.
(Iván Peña Nesbit/AVIAMIL)

Mirage 50EV 3373 takes off from Base Aérea El Libertador for Ecuador on 13 December 2009 after 16 years of service with the FAV.
(Iván Peña Nesbit/AVIAMIL)

Atlas Cheetah C

The Cheetah story

Like Israel before it, in 1981 the Republic of South Africa experienced the imposition of a major French arms embargo (in this case based on UN Security Council Resolution 418). At the time, the South African Air Force (SAAF) had already acquired a total of 58 Mirage IIIC/D/E/Rs and 48 Mirage F.1AZ/CZs. In contrast to Israel, South Africa could not call upon direct US support, nor could it continue to develop its indigenous aerospace industry (headed by Atlas Aircraft Corporation, established in 1965, and later Denel Aviation) with clandestine French support. However, by the time the SAAF issued a requirement for a modern fighter and strike aircraft in the early 1980s, the country was well connected with Israel. The resulting fighter jet, named Cheetah, emerged as a very unusual mix of Israeli, French, and domestically developed technologies.

The first batch of Cheetahs consisted of 16 former Mirage IIIEZ, 16 Mirage IIIDZ/D2Z and a single Mirage IIIRZ airframe, overhauled, rebuilt and provided with a number of Israeli modifications, as well as various new avionics systems. The resulting Cheetah E/D/Rs received canards similar to those of the Kfir, two new hardpoints, an in-flight refuelling probe, new ejection seats and a new nav/attack system. They remained in service for only six years, pending the introduction of a much more capable variant, the Cheetah C.

Much of the background of the Cheetah C remains classified to this day. Suffice to say, in terms of combat capability, the Cheetah C is probably the ultimate variant of the classic delta-winged fighter. The reason for these advanced capabilities is likely to be found in the level of cooperation between Atlas and IAI. Persistent rumours, supported by the fact that after the Cheetah E/D/R projects the SAAF was left without airframes that could be upgraded to a more advanced standard, suggest that South Africa purchased up to 60 Kfir C2/TC2 and Mirage 5/5D airframes from Israel and France, respectively. This enabled a much more comprehensive project to be embarked upon, resulting in the conversion of 38 single-seat Cheetah C and 16 two-seat Cheetah D aircraft. These incorporated not only all the modifications from the earlier design, but also a much more streamlined nose containing the Elta EL/M-2032 radar. The Elbit head-up display was also linked to various other avionics systems (including South African SPJ 200 internal self-protection jammer and RWS 200 radar homing and warning system) via a Mil Std 1553B databus. The shorter but wider intakes were better suited to the more powerful Atar 9K50C-11 engine, and the aircraft was provided with a larger ventral fuel tank with integrated CFD 200 chaff and flare dispensers at its rear end. Although rendering the aircraft slightly slower in terms of maximum level speed and acceleration, the aerodynamic refinements significantly increased the turn rate and reduced the minimum airspeed of the Cheetah C.

The Cheetah C is compatible with a wide range of indigenous South African weapons, including V3C Darter (a much improved version of the V3B Kukri, itself a copy of the R.550 Magic 1), V3S Snake (Rafael Python III), and V4 R-Darter and U-Darter air-to-air missiles, the latter two weapons originally developed by Kentron (later part of Denel Corporation). While the U-Darter was developed as an upgrade of the V3C, the R-Darter is very similar in outward appearance to the Derby, a medium-range, active radar homing missile of Israeli origin. The Cheetah can also carry a range of advanced air-to-ground weapons, including various laser- and TV-guided bombs, the Kentron-developed MUPSOW standoff weapon, and the Torgos, an improved version of the MUPSOW with GPS and a range of guidance options.

The SAAF retired its entire Cheetah C/D fleet in 2008, pending the delivery of Saab JAS 39 Gripen fighters from Sweden.

Ecuador

One of 12 aircraft earmarked for Ecuador, Cheetah D 860 will be used as a spare parts source. It is seen still in SAAF markings at Waterkloof in early 2007. (Jaco du Plessis)

At the same time as negotiating with Venezuela for Mirage 50s and related engine spares, the FAE began to look for additional second-hand aircraft in order to maintain the credibility of its air power following the retirement of the SEPECAT Jaguar. Offers were received from Spain (Mirage F.1), Chile (Mirage 50 Pantera) and South Africa (Atlas Cheetah C).

Following an in-depth analysis, the Atlas Cheetah C was judged the most promising option and in October 2009 Ecuadorian President Rafael Correa announced his intention to acquire 12 examples. Finally, on 30 June 2010 the Ecuadorian Air Force announced the signature of a contract for the purchase of nine Cheetah C fighters with South Africa's state-owned Armscor company. The contract was valued at USD 80 million and included a five-year, renewable maintenance and support package, plus spares and some weapons. The aircraft are expected to replace the Mirage F.1JA/JE within Escuadrón 2112, with deliveries to begin in 2011.

Map of Ecuador

Chapter 6

PERU

Mirage 5P/DP

With Lewis Mejía

By 1965 the Fuerza Aérea del Perú (FAP, Peruvian Air Force) was equipped with material that was showing its age – especially when compared against the modern aircraft then being produced. The FAP still operated the North American F-86F Sabre, Hawker Hunter and examples of the English Electric Canberra bomber, received in 1956.

With Ecuador also looking for new aircraft, and with a long-running border dispute between the two countries, the FAP began to study replacement of its combat fleet. Another factor in the need for modernisation were Peru's periodic disputes with Chile. The Comisión Aérea de Adquisiciones (Acquisitions Air Commission) was created and sent to France, Sweden and Switzerland between February and March 1967 in order to evaluate the Mirage IIIC, Saab 35 Draken and Mirage IIIS, respectively.

Just after the Commission's return, the Six-Day War took place (beginning on 5 June 1967). During the conflict, Israeli pilots scored heavily against Arab fighters, mainly thanks to their 72 Mirage IIICJ jets. In common with many other air forces at the time, the FAP was impressed by the Israeli performance and selected the Mirage IIICJ. By then, however, following a request from Israel, Dassault was developing an attack version of the Mirage. Without radar and with greater range than the Mirage IIICJ, the resulting Mirage 5 caught the attention of the FAP.

Operación Martello

After negotiations, the FAP ordered 14 single-seat Mirage 5s and a pair of two-seaters. The aircraft incorporated some minor modifications requested by the Peruvians, leading to the designation Mirage 5P. The contract was signed under Operación Martello and the aircraft received the serial numbers 182 to 195 (single-seaters), and 196 and 197 for the two-seaters. The contract also included training and weapons, including the AS.30 missile. The AS.30, with a range of around 12km (7 miles), was the first air-to-surface missile in the FAP inventory.

By December 1967, seven pilots had been selected for conversion to the Mirage. The pilots were sent to France between May and July 1968 where they received their

Latin American Mirages

The first Mirage 5P.
(Dassault)

training. Their first stop was Dijon air base, where they received 45 days of ground instruction that took in the aircraft's engineering, systems and procedures. Subsequently they went to Luxeuil air base, where the pilots received instruction on a flight simulator before moving to Mont-de-Marsan for the initial training flights on the Mirage IIIB. At this time, the Armée de l'Air inventory did not yet include the Mirage 5.

The live weapons training phase took place at Cazaux air base. Two pilots (Major Augusto Romero-Lovo Ferrecio and Captain César Gonzalo Luzza) were designated as instructors, in order to prepare future Mirage pilots after their return to Peru.

Meanwhile, on 7 May the first two aircraft were delivered to the FAP at Bordeaux. Two months later they arrived by ship at the port of Salaverry in the region of Lambayeque, The jets were taken immediately to Base Aérea Capitán Quiñónez at Chiclayo. Here, the Mirages were used to equip Escuadrón de Caza 611 'Los Gallos de Pelea', which was previously equipped with Hunters.

Once assembly was complete, on 16 July 1968, Major Romero-Lovo made the first flight of a Mirage 5P in Peru. Captain Gonzalo Luzza recorded a second Mirage flight on the same day. Seven days later, the first public appearance of the Mirage took place over Lima, at Base Aérea Las Palmas, during celebrations for the anniversary of the FAP. The new supersonic jet impressed Peruvian President Fernando Belaunde Terry and other high-ranking officials. The flight from Chiclayo to Lima took place on 20 July and was conducted at supersonic speed. The Mirage recorded a speed of Mach 1.8 during the transit, flying the almost 800km (497 miles) in only 38 minutes.

The remaining 14 Mirages were delivered in pairs, and these arrived every two or three months until December 1969, when the last two examples arrived in Peru. By then, Escuadrón 611 had a flight simulator and all the facilities required to train the

A Mirage 5P with unguided rocket pods.
(Archive Sergio de la Puente)

A Mirage escorts the Peruvian Douglas DC-8 presidential aircraft in the early 1980s. (Archive Amaru Tincopa)

pilots, making it possible to attain a basic operational capability by the end of the same year.

By the beginning of the 1970s, a training programme had been developed for the Mirage 5P pilots and the operational level improved considerably. The Mirages shared the task of defending the border with Colombia and Ecuador together with the last remaining Hunters. The Hunters had been transferred to Base Aérea Capitán Montes, at Talara, in the north of the country.

The first accident took place on 8 March 1971 when FAP 188 was lost. Meanwhile, on 12 May, after an intense training programme, a first AS.20 missile was launched, without a warhead, at the Reque gunnery range. After the impact of the unarmed weapon had been verified, the pilot, Gonzalo Luzza, performed the first firing of a live AS.30 missile, destroying the target.

Progress was also made in terms of maintenance, both on the flight line and at the intermediate level. Escuadrón de Mantenimiento 606 and Escuadrón de Electrónica 605, the first unit of its kind in the FAP, carried out maintenance. Later, the Servicio de Mantenimiento (SEMAN) implemented a major maintenance programme for the aircraft and its major subsystems, including the Atar 9C engine.

A formation of Peruvian Mirage 5Ps is lead by a Mirage 5DP in the early 1970s. (Archive Lewis Mejía)

A Mirage 5P armed with one AS.30 air-to-surface missile. (Archive Amaru Tincopa)

More Mirages

By April 1971, Mirage 5DP serial number 197 had been lost in an accident and the FAP decided it was necessary to replace it. There were also tentative plans to establish a second Mirage squadron. As a result, in 1974 eight single-seat Mirage 5P2s and a single two-seat Mirage 5DP2 were purchased. An option was taken out on another 12 examples, and this was later exercised.

By 1974 the first examples of the second batch of Mirages had begun to arrive and these were used to form Escuadrón 612 at the same base. Deliveries were completed by 1976, with a total of 23 aircraft in both squadrons. After some years of extensive use, both units reached an optimum operational level, and deployments could be made across Peru, including to Base Aérea La Joya in the southern desert.

In 1976 a third purchase took place, exercising some of the options included in the previous contract. One Mirage 5DP2 and seven Mirage 5P3s were received, the last of these configured as interceptors, with the Cyrano IV radar and a new Litton LN-33 inertial navigation system (INS).

In order to revive an earlier project to establish a third Mirage squadron in the south of the country, the remaining options from the second contract were exercised. These covered four single-seat Mirage 5P4s for attack missions. In addition, a Mirage 5DP4 radar trainer and three additional Mirage 5P3s were obtained as attrition replacements. The first examples arrived in Peru in January 1981 and they would soon receive a baptism of fire over the Cordillera del Cóndor.

One of the first two Mirage 5DPs, FAP 196 was lost on 18 May 1988. (Archive Lewis Mejía)

Mirage 5P in action

Fighting broke out after Peru received reports that Ecuador had occupied Peruvian border posts in the jungle-covered mountains of the Cordillera del Cóndor, on the border between both countries. On 24 January 1981, two Mirage 5Ps were ordered to fly an escort mission to protect a Learjet 25B on a reconnaissance mission. The Mirages were armed with air-to-air missiles. The reconnaissance mission was intended to pinpoint the location of the Ecuadorian troops and confirm their infiltration into Peruvian territory. Photographs taken were to be shown the following day during an emergency meeting of the America States Organization.

Seen in France prior to delivery, Mirage 5DP4 FAP 199 was one of the two-seaters received in 1981.
(Archive Sergio de la Puente)

The images confirmed that Ecuadorian troops had crossed the border and entered Peruvian territory, taking up positions around old and abandoned border posts in the Comaina River zone. The posts had been given the names of Ecuadorian cities. After presenting the information to the America States Organization, Peru sent an ultimatum to Ecuador, requesting its troops to leave the zone.

On 26 January, the General Commander of Ala Aérea No. 1, responsible for security in the northwest sector, and the commanders of Grupos Aéreos 6, 7 and 11 received instructions from Lima to execute air operations in support of a combined infantry and helicopter offensive. The Mirage 5Ps of Grupo 6 received the order to provide permanent air cover over the zone from 05.00 on 28 January until the end of 31 January. The intention was to prevent the enemy troops from receiving any kind of air support and to deny airspace to the Ecuadorian Air Force.

Local air superiority was critical to the progress of the land component, which was supported by Mil Mi-6, Mi-8 and Bell 212 helicopters. In addition, the A-37B ground attack jets of Grupo Aéreo No. 7 were to act decisively over enemy positions. The Mirages were to be used to guide the A-37s to their targets, since the Cessnas were not equipped with inertial navigation equipment at that time. The A-37s were to fly from their base at Piura, on the coast, to Comaina, in the mountains.

Grupo Aéreo No. 6 deployed Escuadrón 611 to direct the attack aircraft and to prepare for offensive missions. Meanwhile, Escuadrón 612 was to provide air cover. It was decided to provide three flights of four aircraft each. At any time, one would be operating over the theatre of operations, one would be en route, and one would be returning to Chiclayo. In addition, 18 aircraft were to be maintained on five-minute alert, configured for attack missions, with the pilots in their cockpits.

In order to augment the front-line aircrew, a number of reservist officers, who had flown the Mirage 5 in the past, and continued training on the type at weekends, were sent to Chiclayo. The technicians made great efforts, recording turnaround times of just 15 minutes on aircraft returning from missions.

After 14.00 on 25 January a Learjet and a Canberra of Grupo Aéreo No. 9 from Pisco, each escorted by a pair of Mirage 5Ps from Escuadrón 611, were sent to photograph Ecuadorian positions and surrounding territory. After a very long flight, and when it was completely dark, the Canberra began its return to Peruvian territory. Using an unexpected route, the Canberra alerted the defensive system of Grupo Aéreo No. 11 at Talara, which immediately prepared its Pechora (SA-3 Goa) SAM launchers.

A Peruvian Mirage 5 ready for take-off.
(Archive Lewis Mejía)

Meanwhile, two Su-22Ms were scrambled to shoot down the intruder. Fortunately, a careful visual identification by the Su-22M pilots saved the life of the Canberra crew.

Meanwhile, the Learjet, which had taken off at 16.00, had to wait for improved weather conditions over the target. Just when everything seemed to be proceeding as planned, one of the Mirages reported a failure of its navigation system. The Mirage fell out of formation, and the pilot had to navigate to Chiclayo by dead reckoning before descending for a visual night flight over the sea. With the red light illuminated to indicate low fuel, and after ejecting the external tanks, the aircraft made a barrier landing, suffering minor damage.

28 January was the most intense day of the conflict, during which Grupo 6 completed 40 missions with the Mirage 5, recording 60 flight hours. Tensions continued on the following day, and 30 missions took place, for 47.7 flying hours. At 15.00 two Mirages, flying over the Cordillera del Condor, detected three unidentified helicopters trying to escape to Ecuador at high speed. They descended to assume a position to engage, and identified the intruders while they were crossing the border. However, the Mirage pilots were ordered not to cross the border.

At 13.00 on 31 January another tense situation unfolded when an Ecuadorian SEPECAT Jaguar flew fast over the conflict area. The eyes of the FAP and their Mirages were on a presidential flight, returning to Lima from the conflict zone. Only two Mirage 5s were in the border area and immediately gave chase, but they did not engage the Jaguar. On the fourth day of operations, the Peruvian government declared that they had evicted the enemy and that the Peruvian flag was raised again on the border posts. Between 22 January and 13 March, the Mirage 5 flew a total of 97 missions, comprising 153 flying hours.

Mirage 5P3 FAP 110 with eight Mk 82 bombs with Mk 15 Snakeye retarding fins. Note the Grupo 6 insignia on the fin. (Archive Lewis Mejía)

Aid to Argentina

When Argentina and Chile almost went to war in 1978, the Peruvian government gave its complete support to Argentina, as a result of their own longstanding border disputes with Chile. In case Argentina went to war with Chile, Peru was completely committed to support the former, and FAP A-37Bs were already deployed to Argentina. In addition, it was agreed to sell Argentina 10 Mirage 5P fighters for 1 million US Dollars each, to augment the Argentine Air Force fleet. Once the crisis had ended, negotiations were subject to delay. In 1981 a contract was signed covering the sale of 10 aircraft, with deliveries by the end of 1982 or 1983. When war broke out between Argentina and UK over the Malvinas/Falklands Islands in April 1982, Peru again offered its complete support to Argentina.

Peru initially offered to send a number of combat units to take part in the war, but Argentina refused. Instead, a rapid delivery of the Mirages was negotiated. On 4 June the 10 jets departed Grupo Aéreo No. 6 and, after a stop at Jujuy in northern Argentina, they landed at VI Brigada Aérea of the Argentine Air Force. The Peruvian pilots offered to continue to the south and volunteered to fly combat missions, but the Argentines declined the offer. With the transfer of its aircraft to Argentina, Escuadrón 612 was deactivated on 30 June 1982.

Soviet equipment

In the following years, the FAP Mirages received some Soviet equipment, including an IFF, which was purchased together with the Su-22M. The Mirage was also adapted to carry the R-3 (AA-2 Atoll) air-to-air missile and certain Soviet-built bombs.

In March 1982, Dassault was announced as winner of the Mirage 5P upgrade programme. This was to include installation of a head-up display, a laser telemeter, an inertial navigation system and a refuelling probe. The Mirages also received the capability to launch the Matra R.550 Magic air-to-air missile, especially important for the aircraft equipped with the Cyrano IV radar. The ability to launch the AS.30 missile was retained. Almost all work was conducted by SEMAN at its workshops in Las Palmas and the upgrade was completed in the early 1990s.

Latin American Mirages

A rare photo of a Mirage 5P armed with an R-3 missile, originally received with the Su-22M but later tested on the Mirage. The task of mating the R-3S to the Mirage proved a relatively simple one, since the Mirage 5Ps were wired for US-made AIM-9B Sidewinder AAMs before delivery. The R-3S was developed from the AIM-9B and the installation for both weapons was practically identical.
(Archive César Cruz)

FAP 192 together with a Mirage 2000P and a Canberra. The Mirage 5P is armed with R-3 missiles and bombs.
(Archive Lewis Mejía)

War over the Cenepa

On 17 July 1993, the 25th anniversary of the Mirage 5 was celebrated with a major flight demonstration. By then, however, the units were feeling the effects of the continuous budget cuts imposed on the FAP. By the end of 1994 the situation in Escuadrón 611 was critical. Operations were suspended for five months, after an accident in August, when a two-seater suffered a fuel system failure, killing both pilots.

It was under these circumstances that a new conflict with Ecuador began in late January 1995. The squadron now only had one Mirage 5DP4 on the flight line, this being used by Major Javier Gamboa, the head of the squadron, to train other crews. On 27 January, Captains Patrnogic, Rodríguez and Gambarini were declared ready to fly the Mirage again, while three Mirage 5P4s were returned to service.

In the event, operations over the Cenepa were performed by other units, and it was decided that the Mirage 5P would only be used if the conflict expanded into an open war.

A fragile period of peace existed in the following months, with some minor ground clashes and air alerts. Meanwhile the Mirage 5s were being overhauled and returned to service. By 1996 the aircraft had received chaff and flare launchers and radar warning receivers purchased in Israel. They also received the capability to launch IAI Lizard laser-guided bombs.

Mirage 5DP2 FAP 198 testing the French SAMP EU2 500lb (227kg) laser-guided bomb. (Archive Sergio de la Puente)

On 10 January 1997, the arrival of the Mikoyan MiG-29 fighter saw Escuadrón 612 reactivated and the Mirages were transferred to La Joya to reactivate Escuadrón Aéreo No. 411 (EA-411). The Mirage's time on the unit was short, and in May the aircraft were taken to SEMAN to be prepared for a return to service with Grupo Aéreo No. 6. However, this plan was never realised.

The Mirage 5s operated for a period as a detachment of the Operations Command, seconded to SEMAN, and were flown from Las Palmas and Pisco. With the change of government in 2001, the defence budget was reduced even further, and the FAP was obliged to ground most of its older and more expensive weapons systems, including the Mirage 5s.

A Mirage 5P in maintenance, with black-painted nose. (Archive Amaru Tincopa)

Latin American Mirages

Operational flights came to an end on 21 October 2001, when commanders Rodríguez and Patrnogic, together with Captain Asenio, took a Mirage 5DP4 and a Mirage 5P4 from Pisco to Las Palmas.

Around 10 of the remaining aircraft were preserved at SEMAN, waiting for an order to return to service. A possible sale to another country was also discussed. Finally, on 14 June 2008, a presidential decree ordered the final retirement of the Mirage 5.

The second FAP 194 was a Mirage 5P4 received in 1981. At the time of writing the aircraft was stored at SEMAN. (Archive Lewis Mejía)

182

Mirage 2000P/DP

After the conflict with Ecuador in 1981, the FAP saw it was necessary to upgrade its combat fleet, and began looking for a new fighter. With the approval of Peruvian President Fernando Belaunde Terry, the Consejo de Defensa Nacional (National Defence Council), began to study the aircraft currently available on the market.

The FAP launched an international competition in order to acquire a multi-role fighter, and offers were received for the General Dynamics F-16 Fighting Falcon, the IAI Kfir C7, the MiG-23 and the Dassault Mirage 2000. Only the F-16 and the Mirage 2000 were taken into consideration since the FAP requirement specified a fighter with fly-by-wire controls. The Mirage 2000 was declared the winner since it offered the easiest transition for pilots and technicians accustomed to the Mirage 5P.

In December 1982 the FAP signed the Jupiter I contract, worth 750 million US Dollars, and covering 16 Mirage 2000s (14 single-seaters and a pair of two-seaters), plus 1 Dassault Falcon 20F for VIP transport, a flight simulator and access to the Mirage's onboard computer software in order change the weapons configuration and permit the use of different weapons.

As a consequence of the sale of 10 FAP Mirage 5Ps to Argentina, the Peruvian government decided to increase the Mirage 2000 buy to 26. Negotiations for the resultant Jupiter II contract were hard fought, but, fearing the loss of the entire sale to the Soviets, Dassault agreed to sell all the aircraft, training, spares and the simulator for 650 million US Dollars. As a first payment, Peru used the 50 million US Dollars received from Argentina for the 10 Mirage 5Ps.

Construction of the aircraft began at the Martiniac facilities, where the wings were built, and Argenteuil, responsible for the fuselages. The aircraft were then assembled and tested at Mérignac. The M53-P2 engines were built at Evry, south of Paris.

Meanwhile, the FAP began to select the pilots who would convert to the new fighter in France, choosing them from aircrew qualified on the Mirage 5. In May 1985 Major

FAP 054 is unloaded from a ship at Pisco.
(Archive Lewis Mejía)

Presentation of the first Mirage 2000P/DP aircraft at Las Palmas in December 1986. (Archive Lewis Mejía)

Felipe Conde Garay and Captains Donovan Bartolini Martínez, Ricardo Vílchez Raa and Guido Zavalaga Ortigosa flew to Paris and from there to Mont-de-Marsan air base, where they began a six-month ground training course.

Meanwhile, Alan García Pérez won the presidential elections in Peru in 1985. When he took office he declared a drastic reduction of the defence budget. Among the cuts, García decided to cancel the purchase of the Mirage 2000s, however, Dassault had officially delivered two aircraft (a single-seat Mirage 2000P, and a two-seat Mirage 2000DP) during the Paris Air Salon at Le Bourget on 28 June. Furthermore, crew training was progressing, and it was declared impossible to cancel the contract.

On 8 October 1985, the first solo flights with FAP pilots took place, involving Conde Garay, Vilchez Raa and Zavalaga Ortigosa. Bartolini Martínez followed them the next day, but some days later, the order to return to Lima arrived. Meanwhile, the two Mirages already delivered were embargoed until a discussion on the future of the contract could be arranged.

The new negotiation was named Jupiter III and reduced the total number of aircraft to 12, while a congressional investigation into the purchase was begun. Significantly, the revised contract meant the loss of the ECM equipment for the aircraft, the flight simulator, weapons, and the access to the onboard software. Fortunately, the Falcon 20F had already been delivered to the FAP.

The following year, Peru paid the 36 million US Dollars for the definitive contract, and work was resumed.

Escuadrón 412

After all the problems, the first four Mirages were unloaded at Puerto General San Martín in Pisco, on 23 November 1986. The aircraft had arrived three days earlier in the cargo ship *Francoise Billón*. The jets were immediately taken by road to Capitán Renán Elías Olivera air base, by then the headquarters of Grupo Aéreo No. 9, which was equipped with Canberra bombers. French engineers assembled the Mirages with the support of technicians from FAP's Servicio de Mantenimiento (SEMAN).

On 20 December the aircraft were ferried to La Joya (now Coronel Maldonado Begazo air base), headquarters of Grupo Aéreo No. 4. Here, Escuadrón de Caza 412 'Halcones' was established, under the command of Major Conde Garay. The air base is situated in the middle of the desert in the Arequipa region, some 1,000km (621 miles)

A Mirage 2000DP takes off.
(Chris Lofting)

to the south of Lima and a 45-minute drive by car from the Pacific Ocean coast. From here, the aircraft flew to the Escuela de Oficiales (Officers' School) at Las Palmas air base, Lima, for their official presentation some days later.

The first training course began in January 1986, with eight pilots with experience on the Mirage 5 and Su-22 each receiving 100 flying hours. The last aircraft had been delivered by June 1987 and by then the unit had 18 pilots, and could be declared fully operational. Since 1980, Grupo de Aviación No. 4 had also been equipped with 16 single-seat Su-22Ms and 3 two-seat Su-22UMs, operated by Escuadrón 411.

The poor state of the Peruvian economy, with unemployment and high inflation compounded by the struggle with the guerrillas of the Sendero Luminoso (Shining

Mirage 2000P FAP 053 armed with four 500lb (227kg) Lizard LGBs and dummy Magic 2 air-to-air missiles.
(Author's archive)

Latin American Mirages

A Mirage 2000P banks close to La Joya.
(Archive Amaru Tincopa)

Path) group, saw the defence budget constantly reduced. Most of the military budget was focused on equipment needed to fight the guerrillas, while the materiel required for conventional warfare was neglected.

Despite this situation, by 1991 a group of pilots had been sent to Mont-de-Marsan to improve their capabilities using French flight simulators. At the same time, a Sogitec flight and combat simulator arrived at La Joya. Meanwhile, Escuadrón 411 was disbanded and the Su-22s were sent to the north, leaving only the Mirage 2000 in the south.

Twelve pilots were trained on the Mirage 2000 during 1993. As a result, and on the occasion of the 1994 Feria Internacional del Aire y el Espacio (FIDAE), at Santiago de Chile, held in the last week of March, Majors Walter Milenko Vojvodic Vargas and Héctor Jimmy Mosca Sabate performed a demonstration in FAP 063 and FAP 064. During the show, the Mirage 2000P flew together with the Chilean Mirage 50 Pantera and the Argentine FMA IA-63 Pampa.

At war

The reduction in Peru's military budget since 1985, the reorganisation of the armed forces to combat terrorism, and the economic crisis combined to disturb the military balance between Peru and Ecuador. This led to Ecuador's decision to occupy a number of Peruvian border posts in the disputed area of the Cordillera del Condor. Ecuadorian troops took up positions in December 1994, and clashes between troops from both countries took place in early January. The result was the so-called Cenepa Conflict.

Ala Aérea No. 1, intended to provide defence for the north of the country, was by this time equipped with small numbers of A-37, Mirage 5 and Su-22 jets, and requested

A Mirage 2000P armed with two Magic 2 missiles. (Archive Lewis Mejía)

the support of other units. Unfortunately, Escuadrón 412 possessed only three operational aircraft. After spending 500,000 US Dollars on preparing them for battle, eventually seven Mirage 2000Ps were deployed to the bases of Chiclayo and Talara. From here they were to conduct day and night combat air patrols, escort, reconnaissance and bombing. Although bombing was not a primary role, the Mirages performed two missions during which they dropped 500lb (227kg) Snakeye retarded bombs.

On 10 February two Mirages were flying a CAP over the conflict zone when two Ecuadorian Air Force Mirage F.1 fighters shot down two Su-22s. Immediately the Mirage 2000s were sent to intercept the Mirage F.1s, but when they made radar contact, the attackers detected the emissions on their RWRs and escaped at full speed. Since Peruvian President Alberto Fujimori had given the order not to cross the border, the Mirage 2000s were unable to give chase.

During the war, the Mirage 2000 flew at least 80 combat missions, generating 110 flying hours by day and night.

After the war, the FAP began to purchase new equipment, with MiG-29 and Su-25 aircraft arriving from Belarus. The upgrade of the Mirages was delayed until 1997, when they were modified to launch Lizard laser-guided bombs purchased in Israel for a total of 12 million US Dollars. With the signature of the Itamaraty agreement by Peru and Ecuador, the conflict between the two countries was effectively resolved, and the definitive border was marked.

Among the weapons used by the Mirage 2000P/DP are the internal DEFA 554 30mm cannon, Matra R.550 Magic 2 air-to-air missiles, 250kg (551lb) SAMP 25FI and 500lb (227kg) Mk 82 bombs, Expal BME-330 cluster bombs and BAP-100 anti-runway bombs. Since 1997 the Mirages have also deployed the 500lb Elbit Lizard laser-guided bomb. The aircraft can carry the Thompson-CSF Remora and Caiman electronic countermeasures pods, added in 1998, the MBDA Spirale RWR, plus 16 flare and 112 chaff cartridges.

Latin American Mirages

Three MiG-29s and three Mirage 2000Ps in close formation. (Archive César Cruz)

Difficult times

In June 2001 a group of officers from Grupo Aéreo No. 4 was invited to participate in the Paris Air Salon at Le Bourget and to spend two weeks with the Escadron 275 training centre at Orange air base. The group also visited the Thales workshops at Bordeaux, where they were shown the Mirage 2000-5. Meanwhile, in Peru, almost the entire Mirage 2000 fleet was grounded due to a lack of spares, and the training programme was facing serious difficulties.

Mirage 2000DP FAP 193 carrying two AIM-9 Sidewinder training rounds and seen during Exercise Halcón Cóndor, in which FAP aircraft flew against USAF F-16s. (Eduard Cardenas)

The FAP's main fighter in its current paint scheme.
(Amaru Tincopa archive)

By 2003 a small improvement in operational readiness had taken place, thanks to the work of the unit's own personnel and repairs conducted by Escuadrón de Electrónica 405 technicians. The work by Escuadrón de Electrónica 405 on the multimode Doppler radar saw the technicians win the FAP prize for quality and innovation.

These repairs made it possible for the Mirages to take part in the Passex 2004 exercise with US Navy Boeing F/A-18E/F Super Hornets from the carrier USS *Ronald Reagan*. The Mirages also flew dissimilar air combat tactics exercises with the MiG-29s of Grupo Aéreo No. 6 at La Joya in October 2006.

A US Navy F/A-18C Hornet flies with a Mirage 2000DP and a 2000P during the PASSEX 2004 exercise.
(Archive Lewis Mejía)

From 9 to 18 February 2007, three Mirage 2000Ps, under the command of Colonel Jorge Luis Chaparro Pinto, commander of Grupo 4, were deployed to Chiclayo air base for the Halcón Cóndor exercise. The Mirages took part together with two MiG-29s of Grupo 6 and five USAF F-16 Fighting Falcons. This exercise marked the first appearance of the Mirage in its new two-tone grey and light blue camouflage.

Today

In 2010 the dozen Mirage 2000s are at La Joya with Escuadrón de Caza 412, part of Grupo Aéreo 4. After more than 20 years of service the aircraft have flown in excess of 15,000 hours without accident. At the time of writing, most of the aircraft were out of service and awaiting a major overhaul and upgrade. To date, the government has only assigned 140 million US Dollars to the Mirage 2000 – enough to return most of them to service and replenish the spares stock, but inadequate for a full upgrade.

Modernisation to Mirage 2000-5 standard would cost approximately an extra 360 million US Dollars. Since the government is unwilling to authorise such significant expenditure, the FAP is studying a simplified upgrade for its Mirage fleet.

Mirage 2000DP FAP 193 displays its new two-tone light blue/grey colour scheme. (Archive Lewis Mejía)

Map of Peru

Chapter 7

VENEZUELA

Mirage IIIEV and Mirage 5V/DV

By the end of the 1960s, the Fuerza Aérea Venezolana (FAV, Venezuelan Air Force) was equipped mainly with large numbers of North American F-86 Sabres, the survivors of the de Havilland Vampire and Venom fleet, and a smaller number of English Electric Canberras. Although the Canberras were very useful, the Vampire and Venoms were being retired and the Sabres were almost obsolete. As a result, the FAV began to look for a new combat aircraft, and during 1971 signed two contracts, one with Canadair and one with Dassault. The Canadair contract was for 15 CF-5A and 3 CF-5B Freedom Fighters. The Dassault contract covered nine single-seat Mirage IIIEVs, six single-seat Mirage 5Vs and a pair of two-seat Mirage 5DVs. The French fighter was selected in preference to the Saab Draken and other types.

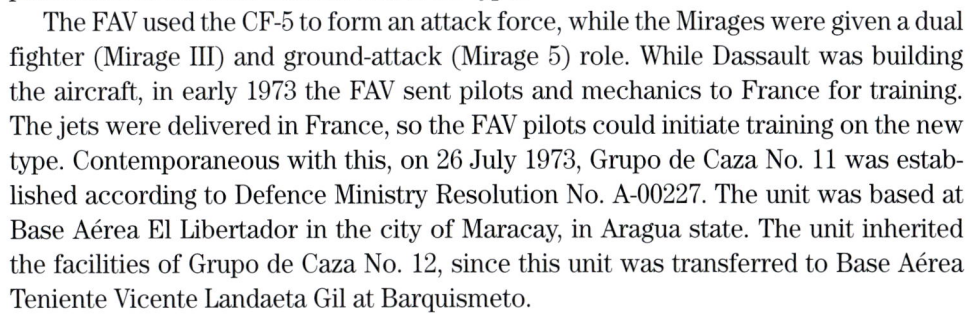

The FAV used the CF-5 to form an attack force, while the Mirages were given a dual fighter (Mirage III) and ground-attack (Mirage 5) role. While Dassault was building the aircraft, in early 1973 the FAV sent pilots and mechanics to France for training. The jets were delivered in France, so the FAV pilots could initiate training on the new type. Contemporaneous with this, on 26 July 1973, Grupo de Caza No. 11 was established according to Defence Ministry Resolution No. A-00227. The unit was based at Base Aérea El Libertador in the city of Maracay, in Aragua state. The unit inherited the facilities of Grupo de Caza No. 12, since this unit was transferred to Base Aérea Teniente Vicente Landaeta Gil at Barquismeto.

The Mirage base was located very close to Caracas, the capital of Venezuela, the Caribbean coast and the Venezuelan Gulf, where the country has its main oil reserves. Grupo No. 11 consisted of Escuadrón 33 'Halcones' with the Mirage III, and Escuadrón 34 'Caciques' with the Mirage 5. Also part of the Grupo was Escuadrón 117, responsible for maintenance.

In 1973 the aircraft arrived in Venezuela and training began. Early operations demonstrated the great advantage of the Mirages over previous equipment and also against the CF-5A. In particular, the Mirage III was prized on account of its Cyrano IIB radar and ability to deploy the Matra R.530 air-to-air missile. In order to increase their firepower, the Mirages were adapted to carry the AIM-9B Sidewinder air-to-air missile, 100 examples of which were received together with the CF-5 in 1972.

On 22 September 1976 the Mirage suffered its first accident in FAV service, when a two-seater was lost, both crew managing to eject. The aircraft lost was FAV 7381, however, it is worth noting here that the serial numbers of the Venezuelan aircraft were assigned in random order, and are not correlative.

Generals Roberto Gruber, Angel González, Carlos Pinaud Arcila, Torres Uribe and Viana Lamas posing in front of a Mirage 5V in 1987.
(FAV)

This accident was followed on 17 September 1979 by the loss of a Mirage IIIEV. Replacements for both aircraft were purchased some time later.

The fleet diminishes

In the early 1980s, Matra R.550 Magic air-to-air missiles were purchased to replace the Sidewinders, and by 1981 the Mirages had been scrambled many times after a number of reports from Venezuelan airliner crews pointing to the presence of Cuban MiG-23 fighters operating from Guyana. The threat posed to Venezuelan oil resources by a possible Cuban attack led to the FAV's decision to buy 24 General Dynamics F-16 Fighting Falcons in 1982. The F-16 was selected in favour of the Mirage 50, Mirage 2000 and the IAI Kfir. Thereafter, the FAV Mirages remained in service as a secondary line of defence.

Meanwhile, accidents continued during the 1980s, and by 1990 the Mirage IIIs 0240, 0624, 2843, 7712, 9325 and 3039, the Mirage 5V 9510 and the Mirage 5DV 5471 had all been lost. This left an operational fleet of only three Mirage IIIs, three Mirage 5s and two Mirage 5DVs. Three of the Mirage IIIs were lost on 5 August 1986, when a flight of four became disorientated during instrument flying and three ran out of fuel. The fourth disobeyed the orders of the flight commander and left the formation, managing to return safely to base.

As a result of accumulated attrition, the squadrons were left with only a limited operational capability. However, and since the aircraft were still relatively modern and were in good condition, it was decided to upgrade the survivors. A contract was signed with Dassault worth 300 million US Dollars. This would see the French company upgrade the older aircraft to Mirage 50 standard. The one exception to the upgrade programme was Mirage IIIEV 1225, which was out of service and in poor condition.

Venezuela

Venezuelan Mirage 5s undergo maintenance.
(FAV)

A Mirage IIIEV before being upgraded to Mirage 50EV standard.
(Author's archive)

Mirage IIIEV FAV 0624 taxies in front of two hardened aircraft shelters.
(HAS)

Latin American Mirages

A Mirage IIIEV armed with two AIM-9B Sidewinder missiles. (FAV)

A formation of F-16A, Mirage IIIEV and two VF-5Bs. (FAV)

Venezuela

A Venezuelan Mirage 5DV formates with an Argentine Mirage IIIEA.
(Dassault)

A Mirage 5V before delivery.
(Dassault)

FAV 5706 in the late 1980s. The aircraft is armed with two Magic and one AM.39 Exocet missiles and 250lb (113kg) bombs. This aircraft was later modified as a two-seater in France, receiving an entirely new forward fuselage.
(Elio Viroli)

Mirage 50EV/DV

The upgrade contract signed by the FAV with Dassault also included the purchase of three Mirage 5M fighters that had already been built for Zaire, but had not been delivered, as the buyer had not paid for them. These three jets would also be upgraded to Mirage 50 standard. Finally, the construction of six Mirage 50EVs and one Mirage 50DV was ordered, these comprising the final first-generation Mirages built. Also, in order to increase the fleet of trainers, Mirage 5V 5706 was modified to a two-seater, receiving a completely new forward fuselage.

The Mirage 50EV was equipped with canard foreplanes, the Atar 9K50 engine that offered increased power and reduced fuel consumption, an aerial refuelling probe, a Thompson-CSF Cyrano IV M3 radar with air-to-air and air-to-surface modes, HOTAS controls, Sherloc RWR, AN/ALE-40 chaff and flares dispensers, a Thompson-CSF VE-110 HUD, INS ULISS 81, IFF Mode 3 and the ability to launch Matra R.550 Magic 2 air-to-air and AM.39 Exocet anti-ship missiles. This last capability, made possible thanks to the addition of the new radar, made the Mirages a very useful asset for anti-shipping missions. As such, the Mirages were widely used to protect Venezuelan waters and to support the ships of the Venezuelan Navy. With the adoption of the Magic 2, the older R.530 and previous-generation R.550 Magic 1 were retired.

With the aircraft sent to France for upgrade, Escuadrón 34 remained non-operational, while Escuadrón 33 retained three aircraft to maintain pilot training requirements. These last three aircraft were upgraded locally. On 22 October 1990 the first upgraded aircraft was delivered in Venezuela. The new two-seater arrived in March 1991, and the remaining deliveries followed in the coming months.

In action

On 4 February 1992 an attempted coup d'etat was launched against Venezuelan President Carlos Andrés Pérez. The coup was led by Army and Air Force personnel, under the command of Lieutenant Colonel Hugo Chávez, but the uprising was immediately contained and Chávez imprisoned.

Colonel Roberto Gruber tested the Mirage 50 in France in 1981. Years later, Venezuelan Mirages were modified to this standard, albeit with different avionics, radar and canards. (Author's archive)

Venezuela

Mirage 50EV FAV 6732 in flight over France.
(Dassault)

In November, a new uprising was planned, headed by the FAV. Air Force General Francisco Visconti Osorio prepared the actions that were to start at Base Aérea El Libertador, where only two Mirage 50EVs were operational: 0160 (from the new-build batch) and 2473 (an upgraded Mirage 5). Also available were 3 North American Rockwell OV-10E and six OV-10A Broncos, the F-16s, 5 C-130Hs, 6 Aeritalia G222s, 2 KC-137s, 8 Aérospatiale AS.330 Super Pumas and 12 Bell UH-1Hs. The aircraft were deployed under the cover of Air Force Day, which is celebrated in December.

The new coup attempt took place at 03.30 on 27 November, with the occupation of Base Aérea El Libertador, the Air Force Academy (equipped with North American T-2 Buckeyes and Tucanos) and other military units and radio stations.

During the occupation of the base, two F-16 pilots, Captain Labarca and Lieutenant Vielma, both loyal to the government, managed to escape in two aircraft that were standing on alert, and flew to Base Aérea Teniente Vicente Landaeta Gil at Barquisimeto. Here they joined the CF-5As of Grupo No. 12 and a number of T-2s.

Arrival of the first Mirage 50EV in Venezuela in an Antonov An-124.
(FAV)

Latin American Mirages

FAV 3373 flying over central Venezuela.
(FAV)

The first actions were carried out by two rebel helicopters, one of which was shot down. Then, at 06.15 the two Mirage 50EVs attacked Venezuelan Army units at Fuerte Tiuna, to the west of Caracas. The Mirages inflicted some casualties among loyalist troops who were meeting to suppress the rebellion.

The Mirages returned to their base and took off again some minutes later, bombing the air base at Barquisimeto after 07.00. The attack was flown together with Broncos, and the raiders succeeded in destroying three CF-5As and damaging another five, together with a civil McDonnell Douglas MD-80 airliner damaged.

As they were returning to their base, the Mirages were intercepted by two loyalist F-16s, returning from a combat air patrol over Caracas. The Mirages selected afterburners and made their escape, but the Broncos were unable to evade the F-16s, and two were shot down.

A Mirage 50DV armed with Magic missiles and bombs.
(GAC11)

Another air strike took place over Caracas at 13.00. In this action, the Francisco de Miranda air base at La Carlota was bombed, as was the presidential palace. During the attack, the F-16s appeared again and damaged one Tucano. Anti-aircraft artillery damaged a Bronco, which crashed shortly before landing at its base. At La Carlota, a Roland SAM shot down another Bronco. One F-16 began to pursue one of the two Mirages, but the pilot made his escape at supersonic speed, landing at Aruba, while the other Mirage escaped to Curaçao.

Despite the rebels' efforts, the loyalist F-16s gained the control of Venezuelan airspace. The rebels were hampered by the fact that mostly they were unable to fly any of the F-16s at El Libertador, and that their two Mirages were out of the battle. Meanwhile, the rebels were also struggling on the ground, and by the afternoon it was clear that the coup attempt was failing. At 15.00 the F-16s bombed El Libertador as loyalist troops approached the base. Minutes later, C-130H FAV 2715 took off with the rebels onboard and headed to Peru, putting an end to the rebellion.

Return to peace

The rebellion led to the removal of many officers from Grupo No. 11, while deliveries of the upgraded and new Mirages continued as normal. On 6 April 1993 the loss of the first Mirage 50 took place, when 2056 crashed after a birdstrike while returning from gunnery training. Later, on 14 October, 0160 was lost some 21 miles from Los Frailes, northwest of Margarita Island, during gunnery practice. The pilot, Lieutenant Pacheco Peña, ejected safely.

In 2003, to commemorate the 30th anniversary of Venezuelan Mirage operations, the two-seater 4212 received a special paint scheme on the tailfin. This was retained until the end of the aircraft's career.

Another Mirage was lost on 17 September 2004, when 6732 suffered an engine failure on the approach to land. The pilot, Captain Jesús María Bevilaqua Paulosky, at-

This Mirage 50DV carries Magic 2 missile training rounds and bombs.
(Author's archive)

Two Mirage 50s refuel from a KC-137.
(FAV)

tempted unsuccessfully to restart the engine, and was forced to eject. The aircraft came down close to the runway end. The final accident involving a FAV Mirage 50 took place on 21 June 2007, and involved 5145. On this occasion, Captain Aldao experienced problems on the approach to El Libertador and was forced to eject over Valencia Lake. Sadly, the vegetation in the lake prevented the pilot from reaching the surface and he drowned.

Following this accident the fleet was grounded for a period. Thereafter, the 13 surviving Mirages (10 Mirage 50Vs and 3 Mirage 50DVs) continued to fly sporadically with Escuadrón 33. Meanwhile, Escuadrón 34 was equipped with four Dassault Falcon 20DC aircraft for maritime surveillance. The Mirages stopped flying during 2008.

Mirage 50EV FAV 5145, taxiing at Base Aerea Natal in Brazil, where it was a participant in Exercise Cruzeiro do Sul 2004 (CRUZEX 04).
(Chris Lofting)

Mirage 50DV FAV 7512 at Base Aérea El Libertador.
(Iván Peña Nesbit/AVIAMIL)

In April 2009 Venezuelan President Hugo Chávez announced that the Mirages were to be retired since they were too old. The Aviación Militar Bolivariana de Venezuela (AMBV, Venezuelan Bolivarian Military Aviation, as the FAV had been renamed) was now equipped with the Sukhoi Su-30MK2 multi-role fighter, 12 of which were transferred to Escuadrón 33.

After some time in storage, the Venezuelan government donated six Mirages (three Mirage 50Vs and three Mirage 50DVs) to the Ecuadorian Air Force. After a number of test flights, the first three Mirages took off from El Libertador for Ecuador on 26 October 2009. After a stopover at Panama, the jets landed at Taura air base in Ecuador on 29 October. The three remaining ex-FAV Mirages were delivered to Ecuador in January 2010.

Mirage 50EV FAV 2353 taxies before departing for Ecuador on 13 December 2009. The aircraft had originally been received by Venezuela in 1993.
(Iván Peña Nesbit/AVIAMIL)

Latin American Mirages

The nose of Mirage 50DV FAV 7512.
(Author's archive)

The tail of FAV 7512.
(Author's archive)

204

Venezuela

The tail of Mirage 50DV FAV 4212.
(Author's archive)

Mirage 50 cockpit.
(Iván Peña Nesbit/AVIAMIL)

Latin American Mirages

FAV 4212 ready for another training mission. It is one of six aircraft delivered to Ecuador in late 2010.
(Iván Peña Nesbit/AVIAMIL)

Mirage 50DV FAV 4212 touches down.
(Iván Peña Nesbit/AVIAMIL)

Venezuela

Map of Venezuela

# Appendix I

APPENDIX I

Individual aircraft histories

1. ARGENTINA | Fuerza Aérea Argentina (FAA, Argentine Air Force)

Mirage IIIDA

Serial Number	c/n	Service Entry	Status	Remarks
I-001	1F/1A	05 09 72	w/o 30 03 79	Commodore Rafael Cantisani ejected in error and was killed during training flight. Pilot made an emergency landing. Crashed 30 03 79 at Derqui after engine failure. Pilots Commodore Viola and Captain Jorge Huck ejected.
I-002	2F/2A	05 02 72	In service	Damaged 1991 in accident. Repaired 1997 at Área de Material Río IV.

Mirage IIIEA

Serial Number	c/n	Service Entry	Status	Remarks
I-003	1F/1D	23 09 72	In service	First flight of a Mirage IIIEA in Argentina on 10 01 73.
I-004	2F/2D	01 11 72	wfu	Repaired 10 79 after being hit by 30mm during a gunnery practice.
I-005	3F/3D	01 12 72	wfu	Damaged 08 10 79 by bird strike.
I-006	4F/4D	18 03 73	w/o 03 07 09	First Mirage IIIE to be overhauled in Argentina, at Río IV workshops. Crashed 03 07 09 at Necochea. Pilot ejected.
I-007	5F/5D	13 04 73	In service	
I-008	6F/6D	05 05 73	In service	Damaged 27 05 09 at Tandil when nose gear collapsed.
I-009	7F/7D	20 05 73	w/o 23 03 76	Destroyed 23 03 76 in accident. Pilot 1st Lieutenant Jorge A. García ejected.
I-010	8F/8D	29 06 73	wfu	In storage at VI Brigada Aérea.
I-011	9F/9D	17 07 73	In service	
I-012	10F/10D	27 07 73	In service	Damaged by in flight fire 09 10 79. Repaired in France.
I-013	11F/1HD	11 79	w/o 01 05 07	Crashed 01 05 87 at Tandil. Pilot 1st Lieutenant Marcos Peretti killed.
I-014	12F/2HD	10 79	w/o 25 08 87	Crashed 25 08 87. Pilot Captain Juan Carlos Franchini Allasia killed.
I-015	13F/3HD	10 79	Combat loss 01 05 82	Shot down 01 05 82 by AIM-9L launched from Sea Harrier XZ423 flown by Flight Lieutenant Paul Barton. Pilot 1st Lieutenant Carlos Perona ejected.
I-016	14F/4HD	11 79	w/o 08 10 83	Crashed 08 10 83 at Río Gallegos while performing aerobatics. Pilot Captain Ricardo González.
I-017	15F/5HD	12 79	In service	Received wiring and equipment to use AIM-9L missiles in 2006.
I-018	16F/6HD	01 80	In service	
I-019	17F/7HD	01 80	Combat loss 01 05 82	Shot down 01 05 82 by Argentine AAA close to BAM Puerto Argentino. Pilot Captain Gustavo A. García Cuerva killed.
I-019	17F/7HD	01 80	Combat loss 01 05 82	Shot down 01 05 82 by Argentine AAA close to BAM Puerto Argentino. Pilot Captain Gustavo A. García Cuerva killed.

Mirage IIIBE

Serial Number	c/n	Service Entry	Status	Remarks
I-020	271	04 82	w/o 06 05 94	Modified from Mirage IIIDA before delivery in 12 82. Destroyed 06 05 84 in accident.
I-021	272	07 82	In service	Modified from Mirage IIIDA before delivery in 01 83.

Dagger/ Finger

Serial Number	c/n	Service Entry	Status	Remarks
C-401	S 07	26 11 78	wfu	Finger III. Retained camouflage scheme. wfu.
C-402	S 18	26 11 78	wfu	Finger III. wfu.
C-403	S 16	26 11 78	Combat loss 21 05 82	Shot down 21 05 82 by AIM-9L launched by Sea Harrier ZA190. Pilot Captain Donadille ejected.
C-404	S 12	26 11 78	Combat loss 21 05 82	Shot down 21 05 82 by AIM-9L launched by Sea Harrier ZA190. Pilot Major Piuma ejected.
C-405	S 03	26 11 78	w/o 31 05 94	Converted to Finger II and later Finger IIIA. Aborted take-off during gunnery practice at Tandil 31 05 94. Emergency barrier was retracted after misunderstanding between tower and pilot. Aircraft left the runway and caught fire. Pilot escaped safely.
C-406	S 13	26 11 78	w/o 26 11 79	Crashed 26 11 79 in Buenos Aires province.
C-407	S 26	06 12 78	Combat loss 21 05 82	Shot down 21 05 82 by AIM-9L launched by Sea Harrier ZA175. Pilot 1st Lieutenant Senn ejected.
C-408	S 09	29 01 79	In service	Prototype of Finger I, later Finger IIIA.
C-409	S 10	6 12 78	Combat loss 21 05 82	Shot down 21 05 82 by AIM-9L launched by Sea Harrier XZ455. Pilot 1st Lieutenant Luna ejected.
C-410	S 06	23 11 78	Combat loss 24 05 82	Shot down 24 05 82 by AIM-9L launched by Sea Harrier ZA193. Pilot 1st Lieutenant Carlos J. Castillo killed.
C-411	S 02	28 12 78	wfu 1991	Finger III.wfu in 1991.
C-412	S 49	28 12 78	In service	Finger III
C-413	S 40	23 01 79	w/o 15 10 95	Finger IIIA. Destroyed 15 10 95 in accident close to Mar Chiquita, Buenos Aires province. Pilot ejected.
C-414	S 41	23 01 79	w/o 09 04 91	Finger IIIB. Withdrawn from service 09 04 91 after accident. Retained camouflage scheme.
C-415	S 04	23 01 79	In service	Finger III
C-416	S 23	03 04 79	In service	Finger III
C-417	S 47	29 01 79	In service	Finger III
C-418	S46	03 04 79	w/o 12 06 87	Crashed 12 06 87 near Tandil. Pilot Captain Fernando Robledo ejected.
C-419	S 35	23 12 80	Combat loss 24 05 82	Shot down 24 05 82 by AIM-9L launched by Sea Harrier XZ457. Pilot Major Puga ejected.
C-420	S 38	03 04 79	In service	Converted into Finger IIIA.
C-421	S 45	04 07 79	wfu 1995	Finger III. Retired to Museo Interfuerzas, Santa Romana, San Luis province.
C-422	S 39	03 04 79	In service	Finger IIIA
C-423	S 34	04 07 79	In service	Finger IIIB
C-424	S 17	04 07 79	wfu 2001	Finger III. Retired, preserved as monument in San Julián with serial C 421.
C-425	T 02	04 07 79	w/o 07 10 80	Dagger two-seater. Lost 07 10 80 in accident close to Tandil.
C-426	T 05	04 07 79	In service	Dagger two-seater.
C-427	S 48	15 02 82	w/o 25 10 93	Modified as prototype Finger I in Israel. Later prototype Finger IIIA. Destroyed 25 10 93 by fire at Río Cuarto workshops.

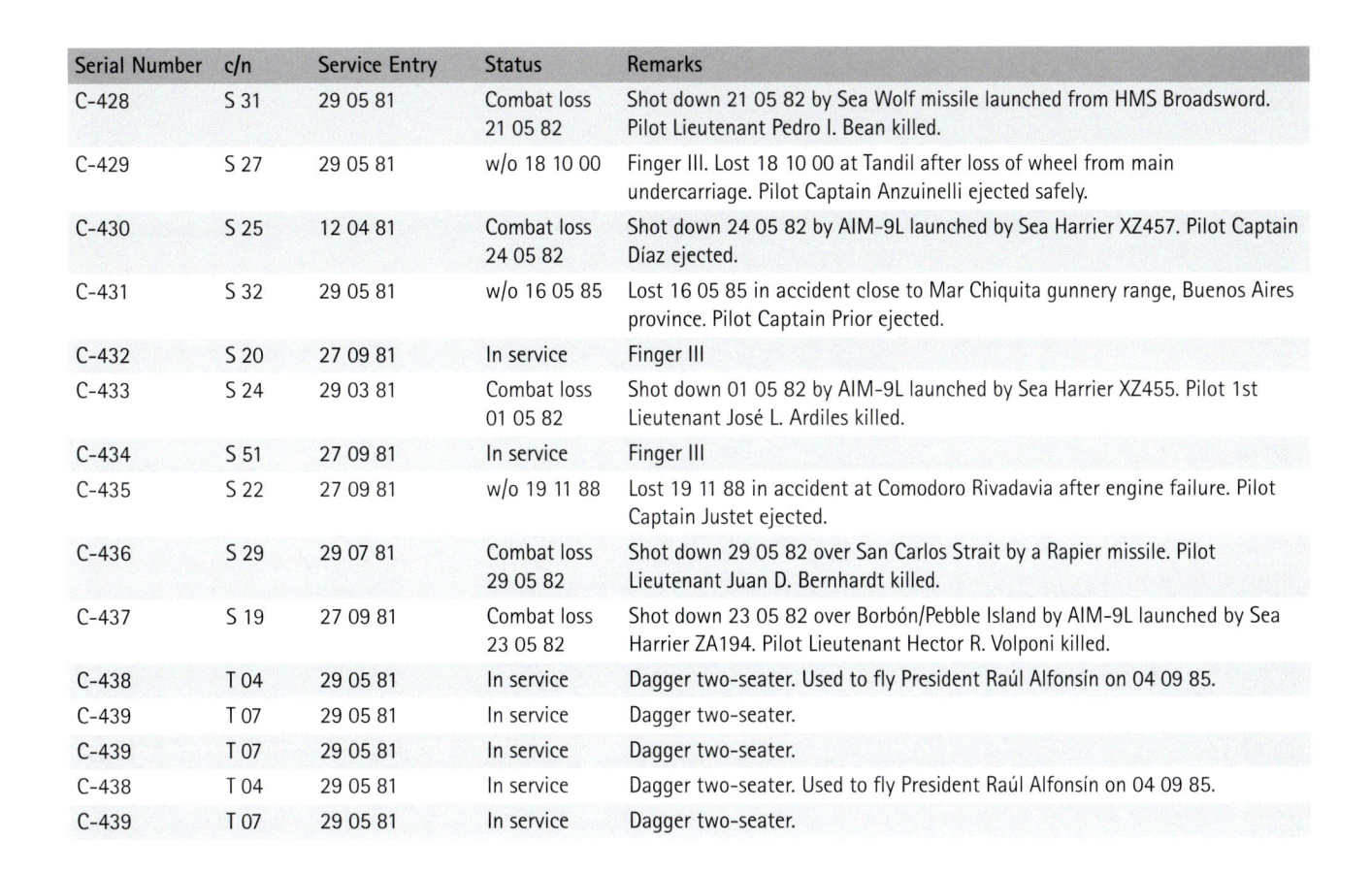

Serial Number	c/n	Service Entry	Status	Remarks
C-428	S 31	29 05 81	Combat loss 21 05 82	Shot down 21 05 82 by Sea Wolf missile launched from HMS Broadsword. Pilot Lieutenant Pedro I. Bean killed.
C-429	S 27	29 05 81	w/o 18 10 00	Finger III. Lost 18 10 00 at Tandil after loss of wheel from main undercarriage. Pilot Captain Anzuinelli ejected safely.
C-430	S 25	12 04 81	Combat loss 24 05 82	Shot down 24 05 82 by AIM-9L launched by Sea Harrier XZ457. Pilot Captain Díaz ejected.
C-431	S 32	29 05 81	w/o 16 05 85	Lost 16 05 85 in accident close to Mar Chiquita gunnery range, Buenos Aires province. Pilot Captain Prior ejected.
C-432	S 20	27 09 81	In service	Finger III
C-433	S 24	29 03 81	Combat loss 01 05 82	Shot down 01 05 82 by AIM-9L launched by Sea Harrier XZ455. Pilot 1st Lieutenant José L. Ardiles killed.
C-434	S 51	27 09 81	In service	Finger III
C-435	S 22	27 09 81	w/o 19 11 88	Lost 19 11 88 in accident at Comodoro Rivadavia after engine failure. Pilot Captain Justet ejected.
C-436	S 29	29 07 81	Combat loss 29 05 82	Shot down 29 05 82 over San Carlos Strait by a Rapier missile. Pilot Lieutenant Juan D. Bernhardt killed.
C-437	S 19	27 09 81	Combat loss 23 05 82	Shot down 23 05 82 over Borbón/Pebble Island by AIM-9L launched by Sea Harrier ZA194. Pilot Lieutenant Hector R. Volponi killed.
C-438	T 04	29 05 81	In service	Dagger two-seater. Used to fly President Raúl Alfonsín on 04 09 85.
C-439	T 07	29 05 81	In service	Dagger two-seater.
C-439	T 07	29 05 81	In service	Dagger two-seater.
C-438	T 04	29 05 81	In service	Dagger two-seater. Used to fly President Raúl Alfonsín on 04 09 85.
C-439	T 07	29 05 81	In service	Dagger two-seater.

Primary differences between Finger variants

Variant	Electronics unit	HUD	ADC	Radar	Doppler	Software	RWR
Finger I	Ferranti	Canadian Marconi, later Thompson CSF	Elta	Elta EL/M 2001B	English	English	No
Finger II	Thompson CSF	Thompson CSF	None	Elta EL/M 2001B	English	Argentine	No
Finger IIIA	Ferranti	Thompson CSF	Elta	Elta EL/M 2001B	English	English	No
Finger IIIB	Thompson CSF	Thompson CSF	Elta	Elta EL/M 2001B	English	Argentine	No
Finger IV	Thompson CSF	Thompson CSF	Elta	Elta EL/M 2001B	English	Argentine	Yes

Mirage 5A Mara

Serial Number	c/n	Service Entry	Status	Remarks
C-403 & C-603	102	04 06 82	In service	Formerly FAP 102. New serial number 603 in 1987. Transferred to Escuadrón X on 13 11 86. Sent to Río IV in 04 88 for Mara modification.
C-404 & C-604	104	04 06 82	w/o 11 06 98	Formerly FAP 104. New serial number 604 in 1987. Transferred to Escuadrón X on 13 11 86. Sent to Río IV in 04 88 for Mara modification. Crashed 11 06 98 at Tandil.
C-407 & C-607	105	04 06 82	w/o 13 03 89	Formerly FAP 105. New serial number 607 in 1987. Transferred to Escuadrón X on 13 11 86. Crashed 13 03 89 at Río Gallegos.
C-409 & C-609	188	04 06 82	w/o 08 08 00	Formerly FAP 188. New serial number 609 in 1987. Transferred to Escuadrón X on 13 11 86. Crashed 08 08 00 at Tandil.
C-410 & C-610	106	04 06 82	In service	Formerly FAP 106. New serial number 610 in 1987. Sent to Río IV in 04 88 for Mara modification.
C-419 & C-619	103	04 06 82	In service	Formerly FAP 103. New serial number 619 in 1987. Modified to Mara.

Serial Number	c/n	Service Entry	Status	Remarks
C-428 & C-628	107	04 06 82	In service	Formerly FAP 107. New serial number 628 in 1987. Transferred to Escuadrón X on March 1988. Modified to Mara.
C-430 & C-630	183	04 06 82	In service	Formerly FAP 183. New serial number 630 in 1987. Transferred to Escuadrón X on 13 11 86. Prototype for Mara project. Re delivered as Mara in 03 88.
C-433 & C-633	185	04 06 82	In service	Formerly FAP 185. New serial number 633 in 1987. Transferred to Escuadrón X on April 1991. Modified to Mara.
C-436 & C-636	186	04 06 82	In service	Formerly FAP 186. New serial number 636 in 1987. Modified to Mara.

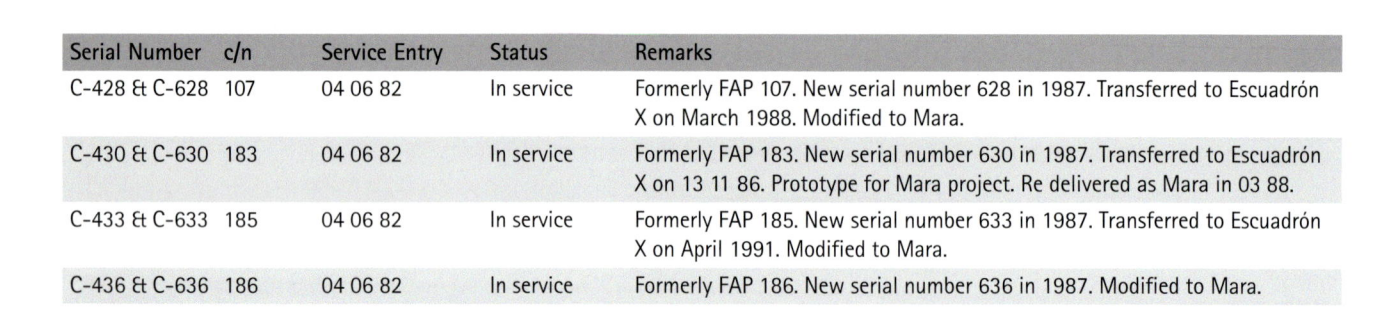

Mirage IIIB/C

Serial Number	c/n	Service Entry	Status	Remarks
C-701	CJ2	27 03 85	ret. 03 94	Built 03 62; 2,803.5 flying hours; final flight 14 12 88. Preserved at Museo Interfuerzas, Santa Romana.
C-702	CJ4	Esc. X 16 3 84, Esc. 55 07 11 86	ret. 03 94	Built 06 61; 2,917 flying hours; final flight 02 02 90. Preserved in San Lorenzo, Santa Fe province.
C-703	CJ12	Esc. X 16 03 84 Esc. 55 07 11 86	ret. 03 94	Built 07 62. In storage at IV Brigada Aérea. Operated at X Brigada Aérea.
C-704	CJ14	Esc. X 16 03 84 Esc. 55 07 11 86	ret. 03 94	Built 04 62; 2,672.45 flying hours; final flight 30 06 89. Operated at X Brigada Aérea. Assigned to Aero Club Dolores. Used in Kfir programme. In storage at IV Brigada Aérea with Mirage VP nose.
C-705	CJ20	1986	27 06 89	Built 5 62. Crashed 27 06 89 at San Juan. Pilot 1st Lieutenant C. Bellini killed.
C-706	CJ22	Esc. X 16 03 84 Esc. 55 07 11 86	ret. 03 94	Built 02 62. Operated at X Brigada Aérea. At Area de Material Quilmes until 2001, later preserved in Buenos Aires.
C-707	CJ29	Esc. X 16 03 84	w/o 29 04 85	Built 10 62; 2,256.15 flying hours. Operated only at X Brigada Aérea. Crashed 29 04 85 at Río Gallegos. Pilot Major A. Kahiara ejected.
C-708	CJ31	17 10 85	ret. 03 94	Built 10 62; 2,243.55 flying hours; final flight 12.08.86. Preserved at Tres Arroyos.
C-709	CJ32	28 10 85	wfu 1991	Built 12 62. At Escuela de Suboficiales de Fuerza Aérea for technical training.
C-710	CJ33	17 10 85	wfu	Built 12 62. In storage at IV Brigada Aérea.
C-711	CJ34	1985	wfu 1991	Built 12 62 as Mirage IIIRJ. Preserved at VI Brigada Aérea.
C-712	CJ40	Esc. X 01 08 85 Esc. 55 07 11 86	ret. 19 06 89	Built 08 63; 2,409 flying hours: 1,949 hours in Israel, 460 in Argentina; final flight 12 06 89. Operated at X Brigada Aérea as replacement for C 707. At Museo Nacional de Aeronáutica since 13 11 92.
C-713	CJ42	1985	To Israel 06 97	Built 02 63; 2,620.02 flying hours; final flight 13 12 90. Used the barrier 21 08 85 after throttle locked on full power. Pilot Lieutenant Maroni. Sold to Israel by file No. 5.408.647 in 06 97, for exhibition at Hatzerim Air Force Museum. Markings of 13 air-to-air kills.
C-714	CJ47	1985	ret. 03 94	Built 04 63. Twelve air to air kills, and one unconfirmed. Preserved at Villa Carlos Paz, Córboba.
C-715	CJ59	1984	Preserved	Built 10 63; Preserved at Liceo Aeronáutico de Funes, Santa Fe, as monument.
C-716	CJ64	1985	ret. 03 94	Built 04 64; 2,093.05 flying hours; final flight 10 05 90. In storage at IV Brigada Aérea.
C-717	CJ65	21 11 85	In service	Built 04 62.In service at Centro de Ensayo de Armamentos y Sistemas Operativos (CEASO) at Río Cuarto.
C-718	CJ66	1986	ret. 03 94	Built 04 63; 2,057.15 flying hours; final flight 28 11 90. In storage at IV Brigada Aérea.
C-719	CJ67	1986	ret. 03 94	Built 04 63; 2,188.55 flying hours; final flight 04 12 90. In storage at IV Brigada Aérea.

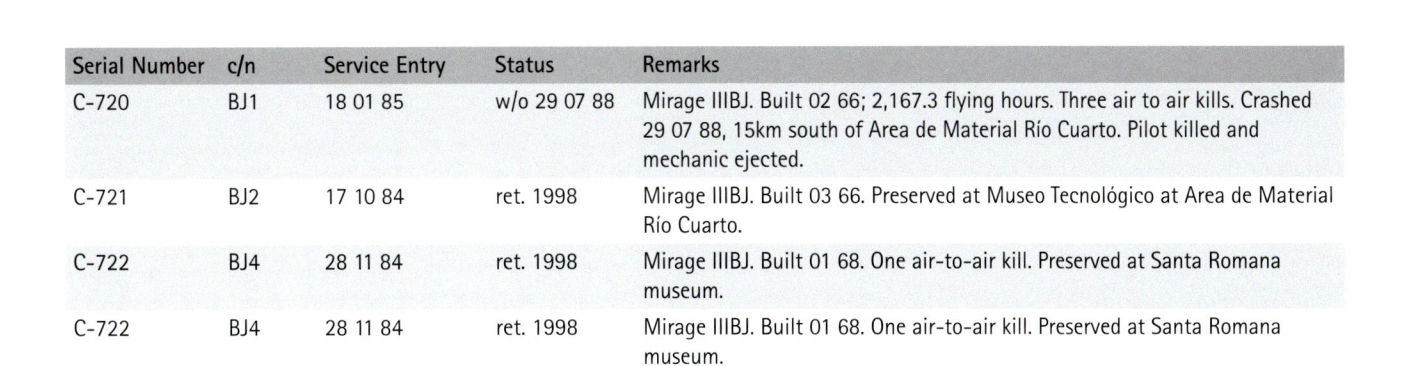

Serial Number	c/n	Service Entry	Status	Remarks
C-720	BJ1	18 01 85	w/o 29 07 88	Mirage IIIBJ. Built 02 66; 2,167.3 flying hours. Three air to air kills. Crashed 29 07 88, 15km south of Area de Material Río Cuarto. Pilot killed and mechanic ejected.
C-721	BJ2	17 10 84	ret. 1998	Mirage IIIBJ. Built 03 66. Preserved at Museo Tecnológico at Area de Material Río Cuarto.
C-722	BJ4	28 11 84	ret. 1998	Mirage IIIBJ. Built 01 68. One air-to-air kill. Preserved at Santa Romana museum.
C-722	BJ4	28 11 84	ret. 1998	Mirage IIIBJ. Built 01 68. One air-to-air kill. Preserved at Santa Romana museum.

2. BRAZIL | Força Aérea Brasiliera (FAB, Brazil Air Force)

Mirage IIIDBR (F-103D)

Serial Number	c/n	Service Entry	Status	Notes
4900	18507	06 04 73	w/o 20 11 80	Crashed at Anápolis 20 11 80. Pilot Teniente Coronel Aviador Mauro Lazzarini de A. Silva killed.
4901	18708 – 2F/2A	06 04 73	ret. 15 12 05	
4902	18709	06 04 73	w/o 18 08 81	Crashed 18 08 81 at Anápolis. Pilot Coronel Aviador Josué Jonas B. M. Falcão killed.
4903	18710	06 04 73	w/o 19 05 82	Crashed 19 05 82 at Natal. Pilot ejected.
4904	18711 BE271F/21A/B	24 02 84	ret. 15 12 05	Formerly Armée de l'Air. Contract 01/DIRMA/83. Named 'Ayrton Senna'.
4905	18712	15 06 84	w/o 25 07 84	Formerly Armée de l'Air. Contract 01/DIRMA/83. Crashed 25 07 84 at Petrolina. Pilot ejected.
4906	DBR6	29 05 89	ret. 15 12 05	Formerly Armée de l'Air. Contract 11/12/DIRMA/87.
4907	DBR5	29 06 89	w/o 15 02 90	Formerly Armée de l'Air. Contract 11/12/DIRMA/87. Crashed 15 02 90 at Anápolis.
4908	M201	1999	ret. 15 12 05	Formerly Zaire Air Force. Built as Mirage 5.
4909	3F/5A	1999	ret. 15 12 05	Formerly Zaire Air Force. Built as Mirage 5.
4909	3F/5A	1999	ret. 15 12 05	Formerly Zaire Air Force. Built as Mirage 5.

Mirage IIIEBR (F-103E)

Serial Number	c/n	Service Entry	Status	Remarks
4910	1F/1A	06 04 73	ret. 15 12 05	Arrived Anápolis 01 10 72. First flight in Brazil 27 03 73. At end of career it received original polished metal scheme.
4911	2F/2A	06 04 73	w/o 28 09 88	Collided 28 09 88 with 4923 at Manaus. Pilot Major Aviador Enrique Roberto Martins killed.
4912	18704 – 3F/3A	06 04 73	w/o 28 06 79	Crashed 28 06 79. Pilot ejected.
4913	4F/4A	06 04 73	ret. 15 12 05	
4914	5F/5A	06 04 73	ret. 15 12 05	
4915	6F/6A	06 04 73	ret. 15 12 05	

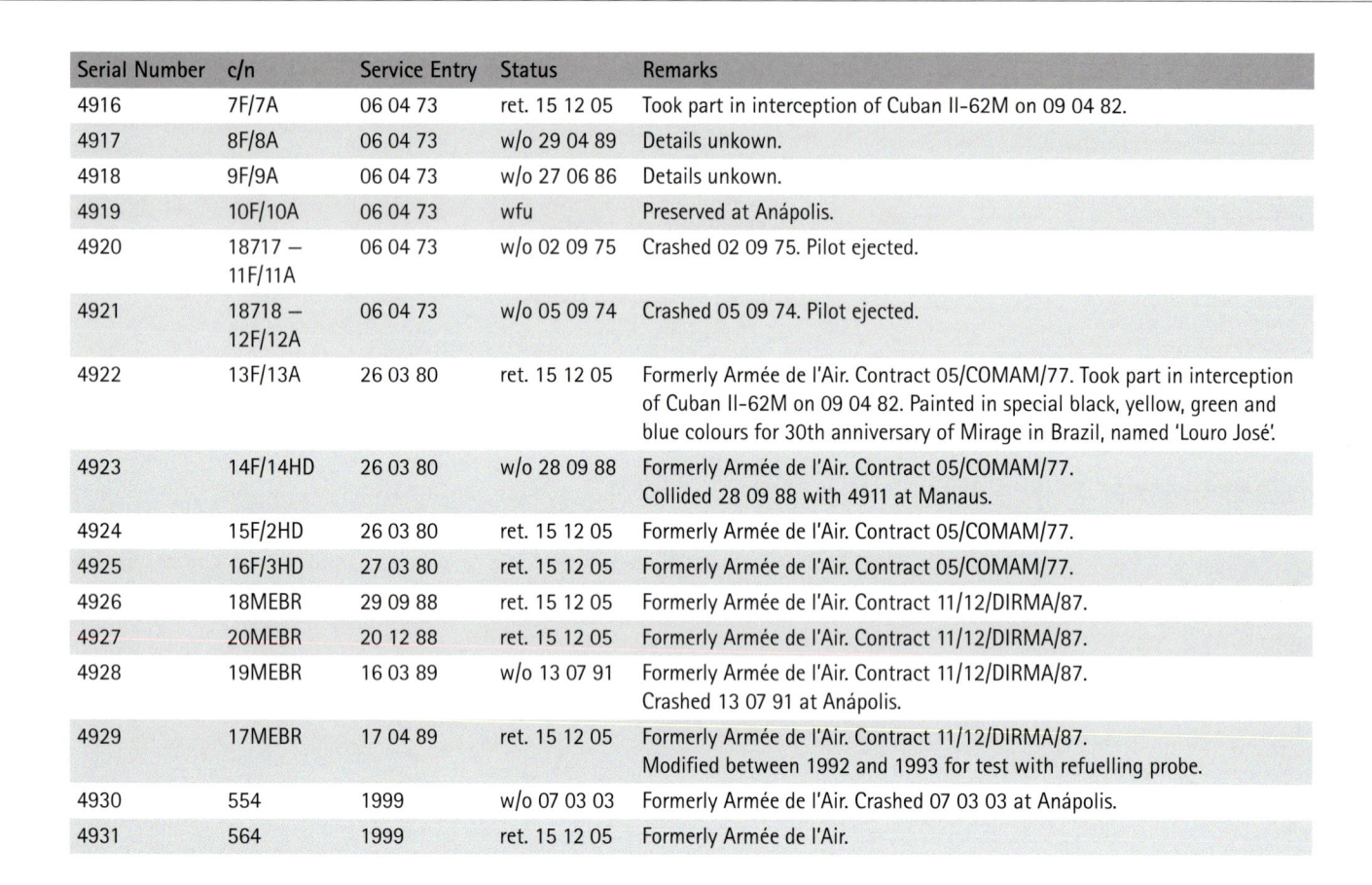

Serial Number	c/n	Service Entry	Status	Remarks
4916	7F/7A	06 04 73	ret. 15 12 05	Took part in interception of Cuban Il-62M on 09 04 82.
4917	8F/8A	06 04 73	w/o 29 04 89	Details unkown.
4918	9F/9A	06 04 73	w/o 27 06 86	Details unkown.
4919	10F/10A	06 04 73	wfu	Preserved at Anápolis.
4920	18717 – 11F/11A	06 04 73	w/o 02 09 75	Crashed 02 09 75. Pilot ejected.
4921	18718 – 12F/12A	06 04 73	w/o 05 09 74	Crashed 05 09 74. Pilot ejected.
4922	13F/13A	26 03 80	ret. 15 12 05	Formerly Armée de l'Air. Contract 05/COMAM/77. Took part in interception of Cuban Il-62M on 09 04 82. Painted in special black, yellow, green and blue colours for 30th anniversary of Mirage in Brazil, named 'Louro José'.
4923	14F/14HD	26 03 80	w/o 28 09 88	Formerly Armée de l'Air. Contract 05/COMAM/77. Collided 28 09 88 with 4911 at Manaus.
4924	15F/2HD	26 03 80	ret. 15 12 05	Formerly Armée de l'Air. Contract 05/COMAM/77.
4925	16F/3HD	27 03 80	ret. 15 12 05	Formerly Armée de l'Air. Contract 05/COMAM/77.
4926	18MEBR	29 09 88	ret. 15 12 05	Formerly Armée de l'Air. Contract 11/12/DIRMA/87.
4927	20MEBR	20 12 88	ret. 15 12 05	Formerly Armée de l'Air. Contract 11/12/DIRMA/87.
4928	19MEBR	16 03 89	w/o 13 07 91	Formerly Armée de l'Air. Contract 11/12/DIRMA/87. Crashed 13 07 91 at Anápolis.
4929	17MEBR	17 04 89	ret. 15 12 05	Formerly Armée de l'Air. Contract 11/12/DIRMA/87. Modified between 1992 and 1993 for test with refuelling probe.
4930	554	1999	w/o 07 03 03	Formerly Armée de l'Air. Crashed 07 03 03 at Anápolis.
4931	564	1999	ret. 15 12 05	Formerly Armée de l'Air.

Mirage 2000C/B (F-2000C/B)

Serial Number	c/n	Service Entry	Status	Remarks
4932	19	04 09 06	In service	Mirage 2000B. Formerly Armée de l'Air.
4933			In service	Mirage 2000B. Formerly Armée de l'Air.
4940	78	04 09 06	In service	Mirage 2000C. Formerly Armée de l'Air.
4941	96	10 06	In service	Mirage 2000C. Formerly Armée de l'Air.
4942	140	10 06	In service	Mirage 2000C. Formerly Armée de l'Air.
4943			In service	Mirage 2000C. Formerly Armée de l'Air.
4944			In service	Mirage 2000C. Formerly Armée de l'Air.
4945			In service	Mirage 2000C. Formerly Armée de l'Air.
4946	13		In service	Mirage 2000C. Formerly Armée de l'Air.
4947			In service	Mirage 2000C. Formerly Armée de l'Air.
4948	154	27 08 08	In service	Mirage 2000C. Formerly Armée de l'Air.
4949		27 08 08	In service	Mirage 2000C. Formerly Armée de l'Air.

3. CHILE | Fuerza Aérea de Chile (FACh, Chile Air Force)

Mirage 50

Serial Number	c/n	Service Entry	Status	Notes
501		15 11 80	ret. 12 07	Mirage 50CF. Formerly Armée de l'Air 1. Modified to Pantera. In storage.
502		15 11 80	ret. 12 07	Mirage 50CF. Formerly Armée de l'Air 3. Modified to Pantera. In storage.
503		15 11 80	ret. 12 07	Mirage 50CF. Formerly Armée de l'Air 5. Modified to Pantera. Preserved at Museo Aeronáutico in Los Cerrillos.
504		15 11 80	ret. 12 07	Mirage 50CF. Formerly Armée de l'Air 8. Modified to Pantera. In storage at El Bosque.
505		15 11 80	ret. 12 07	Mirage 50CF. Formerly Armée de l'Air 16. Modified to Pantera. In storage at El Bosque.
506		15 11 80	ret. 12 07	Mirage 50CF. Formerly Armée de l'Air 23. Modified to Pantera. In storage at El Bosque.
507		15 11 80	ret. 12 07	Mirage 50CF. Formerly Armée de l'Air 28. Modified to Pantera. In storage.
508		15 11 80	ret. 12 07	Mirage 50CF. Formerly Armée de l'Air 30. Modified to Pantera. Used for laser guided bomb tests. In storage.
509		27 04 82	ret. 12 07	Mirage 50C. Modified to Pantera. In storage.
510	510/2F/2HD	27 04 82	ret. 12 07	Mirage 50C. Modified to Pantera. Preserved at Escuela de Aviación, El Bosque.
511		1982	ret. 12 07	Mirage 50C. Modified to Pantera. In storage.
512		1982	w/o 18 06 89	Mirage 50C. Crashed 18 06 89, pilot killed.
513		06 01 83	ret. 12 07	Mirage 50C. Modified to Pantera. In storage.
514	514/6F/6HD	06 01 83	ret. 12 07	Mirage 50C. Prototype for the Bracket and Pantera programmes. In storage.
515		28 05 82	ret. 12 07	Mirage 5D. Modified to Mirage 50DC, later to Pantera. In storage at El Bosque.
516		24 11 82	w/o	Mirage 5D. Modified to Mirage 50DC. Crashed, date unknown, possibly 12 03 93
516		01 06 87	ret. 12 07	Mirage IIIBE. Modified to Mirage 50DC, later to Pantera. Last example upgraded. In storage at El Bosque.

Mirage 5MA/MD Elkan

Serial Number	c/n	Service Entry	Status	Remarks
701	01	1995	Preserved	Mirage 5MA. Formerly Belgian AF BA 01. Preserved at Museo Aeronáutico, Los Cerrillos.
702	04	1995	ret. 27 12 06	Mirage 5MA. Formerly Belgian AF BA 04. In storage.
703	11	1995	ret. 27 12 06	Mirage 5MA. Formerly Belgian AF BA 11. In storage.
704	23	1995	ret. 27 12 06	Mirage 5MA. Formerly Belgian AF BA 23. In storage.
705	37	1995	wfu	Mirage 5MA. Formerly Belgian AF BA 37. In storage at Escuela de Aviación, El Bosque.
706	39	1995	wfu	Mirage 5MA. Formerly Belgian AF BA 39. Used to test refuelling probe. In storage at Escuela de Aviación, El Bosque.
707	46	1995	ret. 27 12 06	Mirage 5MA. Formerly Belgian AF BA 46. In storage.
708	48	1995	wfu	Mirage 5MA. Formerly Belgian AF BA 48. In storage at Escuela de Aviación, El Bosque.
709	50	1995	ret. 27 12 06	Mirage 5MA. Formerly Belgian AF BA 50. In storage.

Serial Number	c/n	Service Entry	Status	Remarks
710	52	1995	ret. 27 12 06	Mirage 5MA. Formerly Belgian AF BA 52. In storage.
711	56	1995		Mirage 5MA. Formerly Belgian AF BA 56. In storage at Escuela de Aviación, El Bosque.
712	57	1995	ret. 27 12 06	Mirage 5MA. Formerly Belgian AF BA 57. In storage.
713	59	1995	ret. 27 12 06	Mirage 5MA. Formerly Belgian AF BA 59. In storage.
714	60	1995	ret. 27 12 06	Mirage 5MA. Formerly Belgian AF BA 60. Prototype for Mirage 5 MirSIP. In storage.
715	62	1995	wfu	Mirage 5MA. Formerly Belgian AF BA 62. Preserved for display at Escuela de Aviación, El Bosque.
716	201	1995	wfu	Mirage 5MD. Formerly Belgian AF BD 01. In storage at Escuela de Aviación, El Bosque.
717	203	1995	wfu	Mirage 5MD. Formerly Belgian AF BD 03. In storage at Escuela de Aviación, El Bosque.
718	204	1995	ret. 27 12 06	Mirage 5MD. Formerly Belgian AF BD 04. In storage.
719	214	1995	ret. 27 12 06	Mirage 5MD. Formerly Belgian AF BD 14. In storage.
720	215	1995	ret. 27 12 06	Mirage 5MD. Formerly Belgian AF BD 15. In storage.
721	313	1995	wfu	Mirage 5BR. Formerly Belgian AF BR 13. Did not receive MirSIP modification. Gate guard at Escuela de Guerra.
722	325	1995	wfu	Mirage 5BR. Formerly Belgian AF BR 25. Did not receive MirSIP modification. Preserved at Museo Aeronáutico at Los Cerrillos.
723	326	1995	wfu	Mirage 5BR. Formerly Belgian AF BR 26. Did not receive MirSIP modification. At Escuela de Especialidades.
724	327	1995	wfu	Mirage 5BR. Formerly Belgian AF BR 27. Did not receive MirSIP modification. At Escuela de Especialidades.
725	212	1995	wfu	Mirage 5BD. Formerly Belgian AF BD 12. Did not receive MirSIP modification. Gate guard at Base Aérea Cerro Moreno, Antofagasta.

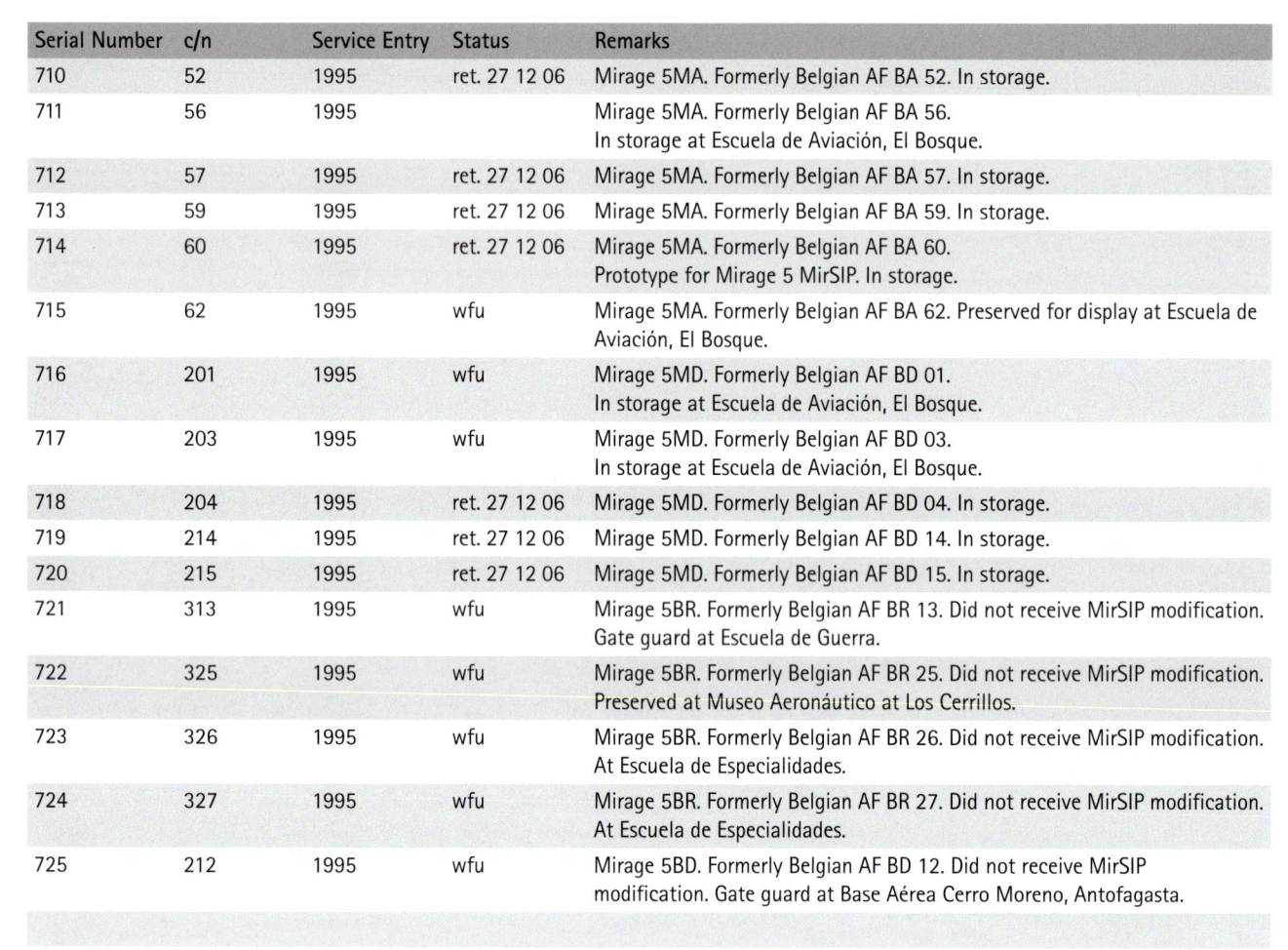

Note: Two Mirage 5s were lost but the FACh has not released the serial numbers. The accidents took place on 07 11 99 and 14 01 02.

4. COLOMBIA | Fuerza Aérea Colombiana (FAC, Colombian Air Force)

Mirage 5COD/COR/COA

Serial Number	c/n	Service Entry	Status	Remarks
FAC 3001	508	13 03 72	In service	Mirage 5COD
FAC 3002	478	20 03 72	In service	Mirage 5COD
FAC 3011 & FAC 3035		19 10 73	In service	Mirage 5COR. Upgraded to Mirage 5COA. New serial number FAC3035, date unknown
FAC 3012		19 11 73	w/o	Mirage 5COR. Lost in accident, date unknown.
FAC 3021		21 03 72	In service	Mirage 5COA
FAC 3022		1972	In service	Mirage 5COA
FAC 3023		17 09 72	w/o 07 04 97	Mirage 5COA. Crashed 07 04 97.
FAC 3024		21 03 72	In service	Mirage 5COA. First example delivered. Conducted first air to air refuelling.
FAC 3025		21 03 72	w/o 12 08 72	Mirage 5COA. Crashed 12 08 72 after tyre blowout on main landing gear. Pilot killed.

Serial Number	c/n	Service Entry	Status	Remarks
FAC 3026		1972	w/o 29 09 98	Mirage 5COA. First example upgraded in Colombia. Crashed 29 09 98.
FAC 3027		1972	In service	Mirage 5COA
FAC 3028		1972	w/o	Mirage 5COA. Lost in accident, date unknown.
FAC 3029		1972	w/o	Mirage 5COA. Prototype for upgrade, modified by IAI in Israel. Lost in accident, date unknown.
FAC 3030		1973	In service	Mirage 5COA
FAC 3031		1973	In service	Mirage 5COA
FAC 3032		1973	w/o 06 85	Mirage 5COA. Crashed 06 85.
FAC 3033		1973	In service	Mirage 5COA
FAC 3034		17 07 73	In service	Mirage 5COA

Kfir C2/ C7/ C12/ TC7/ TC12

Serial Number	c/n	Service Entry	Status	Remarks
FAC 3003	B 08	1990		Kfir TC7. Formerly Israel AF 308. First Kfir TC2 in service in Israel.
FAC 3004	B 04	13 07 09	w/o 20 07 09	Kfir TC2. Formerly Israel AF 304. Crashed 20 07 09 while taking off from Cartagena before delivery. Pilots escaped safely, aircraft destroyed.
FAC 3005	B 05	14 01 10		Kfir TC12. Formerly Israel AF 309.
FAC 3006	B 06			Kfir TC12. Formerly Israel AF ???.
FAC 3007				Kfir TC7. Formerly Israel AF ???.
FAC 3040	54	1989		Kfir C7. Formerly Israel AF 729 & 829.
FAC 3041	58	1989		Kfir C7. Formerly Israel AF 745 & 845.
FAC 3042	59 or 60	1989	w/o 02 05 95	Kfir C7. Crashed 02 05 95.
FAC 3043	61	1989		Kfir C7. Formerly Israel AF 757 & 857.
FAC 3044	62	1989		Kfir C7. Formerly Israel AF 779 & 879.
FAC 3045	75	28 04 89		Kfir C7. Formerly Israel AF 824. First Kfir delivered. Prototype for Kfir C7.
FAC 3046	93	1989	w/o 04 06 03	Kfir C7. Formerly Israel AF 852. Crashed 04 06 03 after engine ingested a bird. Pilot ejected.
FAC 3047		1989		Kfir C7. Formerly Israel AF ???.
FAC 3048	102	1989		Kfir C7. Formerly Israel AF 866.
FAC 3049	104	1989		Kfir C7. Formerly Israel AF 868. Holds speed record in Colombia at Mach 1.98.
FAC 3050	122	1989		Kfir C10. Formerly Israel AF 892.
FAC 3051	123	1989		Kfir C10. Formerly Israel AF 894.
FAC 3052	176	13 07 09		Kfir C12. Formerly Israel AF 523.
FAC 3053	180	13 07 09		Kfir C12. Formerly Israel AF 532.
FAC 3054	193	13 07 09		Kfir C12. Formerly Israel AF 559.
FAC 3055	187	11 01 10		Kfir C10. Formerly Israel AF 543.
FAC 3056	191	11 01 10		Kfir C10. Formerly Israel AF 553.
FAC 3057	198	11 01 10		Kfir C10. Formerly Israel AF 566.
FAC 3058	184			Kfir C10. Formerly Israel AF 570.
FAC 3059		01 06 10		Kfir C10. Formerly Israel AF 561.
FAC 3060	188	01 06 10		Kfir C10. Formerly Israel AF 547.
FAC 3061				Kfir C10. Formerly Israel AF ???.

5. ECUADOR | Fuerza Aérea Ecuatoriana (FAE, Ecuadorian Air Force)

Mirage F.1JE/BE

Serial Number	c/n	Service Entry	Status	Remarks
801		05 79	In service	Mirage F.1JE
802	137	05 79	In service	Mirage F.1JE
803		05 79	In service	Mirage F.1JE
804		05 79	w/o 25 06 80	Mirage F.1JE. Crashed 25 6 80.
805		1979	In service	Mirage F.1JE
806		1979	In service	Mirage F.1JE. Shot down Peruvian Su-22M 10 02 95.
807		1979	In service	Mirage F.1JE. Shot down Peruvian Su-22M 10 02 95.
808	233	1979	In service	Mirage F.1JE
809		1979	In service	Mirage F.1JE
810		1979	w/o 07 03 83	Mirage F.1JE. Crashed 07 03 83.
811		1979	In service	Mirage F.1JE
812		1979	In service	Mirage F.1JE
813		1979	In service	Mirage F.1JE
814	251	1979	In service	Mirage F.1JE
815		1979	w/o 1985	Mirage F.1JE. Crashed 1985.
816		1979	In service	Mirage F.1JE
830		27 06 80	In service	Mirage F.1BE
831		27 06 80	w/o 23 01 88	Mirage F.1BE. Crashed 23 01 88.

Kfir C2/TC2

Serial Number	c/n	Service Entry	Status	Remarks
901	142	25 03 82	In service	Modified to Kfir CE.
902	143	25 03 82	In service	Modified to Kfir CE.
903	144	25 03 82	w/o 12 08 94	Crashed 12 08 94.
904	145	25 03 82	w/o 03 05 89	Crashed 03 05 89.
905	146	25 03 82	In service	Not modified. Shot down Peruvian A-37B on 10 02 95.
906	147	1982	In service	Modified to Kfir CE.
907	148	1982	In service	Not modified.
908	149	1982	In service	Modified to Kfir CE.
909	150	1982	In service	Modified to Kfir CE.
910	151	1982	w/o 07 02 85	Crashed 07 02 85.
911		1996	In service	Not modified.
912		1996	In service	Not modified.
913		1996	w/o 24 04 98	Crashed 24 04 98.
914	131 & 153	1999	In service	Ex Israeli AF 871. Prototype for Kfir 2000.
915	125 & 152	1999	In service	Ex Israeli AF 896. Ground demonstrator for Kfir 2000, modified in Israel by IAI.
930	B 11	25 03 82	In service	
931	B 12	25 03 82	w/o 21 10 04	Crashed 21 10 04.
932	B 02	1996	In service	Ex Israeli AF 302

Mirage 50EV/DV

Serial Number	c/n	Service Entry	Status	Remarks
1297		26 10 09	In service	Mirage 50EV. Formerly FAV 1297.
2353		13 10 09	In service	Mirage 50EV. Formerly FAV 2353.
3373		13 10 09	In service	Mirage 50EV. Formerly FAV 3373.
4212		26 10 09	In service	Mirage 50DV. Formerly FAV 4212.
5706		13 10 09	In service	Mirage 50DV. Formerly FAV 5706.
7512		26 10 09	In service	Mirage 50DV. Formerly FAV 7512.

Cheetah C/D2

Serial Number	c/n	Service Entry	Status	Remarks
				Cheetah C. Former South African Air Force 359. At Denel for overhaul prior to del. to Ecuador, 06 10.
				Cheetah C. Former South African Air Force.
				Cheetah C. Former South African Air Force.
				Cheetah C. Former South African Air Force.
				Cheetah C. Former South African Air Force.
				Cheetah C. Former South African Air Force.
				Cheetah C. Former South African Air Force.
				Cheetah C. Former South African Air Force.
				Cheetah C. Former South African Air Force.
	200			Cheetah D2. Former South African Air Force 844. Crashed in 1993; rebuilt by using rear fuselage of 836; converted as technology demonstrator. At Denel for overhaul prior to del. to Ecuador, 06 10.
				Cheetah D2. Former South African Air Force 845. At Denel for overhaul prior to del. to Ecuador, 06 10.
				Cheetah D2. Former South African Air Force.
	110F			Cheetah B. Former South African Air Force 860. For spare parts only. At Denel for overhaul prior to del. to Ecuador, 06 10.
				Cheetah R. Former South African Air Force 845. For spare parts only. At Denel for overhaul prior to del. to Ecuador, 06 10.
				Cheetah ? Former South African Air Force. For spare parts only.

6. PERU | Fuerza Aérea del Perú (FAP, Peruvian Air Force)

Mirage 5P/5DP

Serial Number	c/n	Service Entry	Status	Remarks
182		1968	Unknown	Mirage 5P
183		1968	To FAA 06 82	Mirage 5P. Del. to FAP 1974; del. to FAA as C 630, 06 82.
184		1968	w/o	Mirage 5P. Crashed, date unknown
184		1981	w/o	Mirage 5P4. Crashed, date unknown
185		1968	to FAA, 06 82	Mirage 5P. Del. to FAP 1974; del. to FAA as C 633, 06 82.
186		1968	to FAA, 06 82	Mirage 5P. Del. to FAP 1974; del. to FAA as C 636, 06 82.
187		1968	w/o	Mirage 5P. Crashed, date unknown
187		1981	wfu 1997	Mirage 5P4. In storage at SEMAN, Las Palmas.
188		1968	w/o 08 03 71	Mirage 5P. Crashed 08 03 71.
188		1974	to FAA 06 82	Mirage 5P2. Delivered to Argentina as C 609, 06 82.
189		1968	w/o	Mirage 5P. Crashed, date unknown
189		1981	w/o 10 03 87	Mirage 5P4. Crashed 10 03 87.
190		1968	w/o	Mirage 5P. Crashed, date unknown
190		1981	Unknown	Mirage 5P3
191		1968	w/o	Mirage 5P. Crashed , date unknown
191		1981	ret. 1997	Mirage 5P3
192		1968	ret. 1997	Mirage 5P. Preserved at Museo Aeronáutico at Las Palmas.
193		1968	w/o	Mirage 5P. Crashed , date unknown
193		1981	w/o 29 08 00	Mirage 5P3. Crashed 29 08 00
194		1968	w/o	Mirage 5P. Crashed , date unknown
194		1981	ret. 1997	Mirage 5P4. In storage at SEMAN, Las Palmas.
195		1968	Unknown	Mirage 5P
196		1968	w/o 18 05 88	Mirage 5DP. Crashed 18 05 88.
197		1968	w/o 09 04 71	Mirage 5DP. Crashed 09 04 71.
197		1976	Unknown	Mirage 5DP2
198		1974	ret. 1997	Mirage 5DP2. In storage at SEMAN, Las Palmas.
199		1981	ret. 1997	Mirage 5DP4. In storage at SEMAN, Las Palmas.
101		1974	w/o 22 03 77	Mirage 5P2. Crashed 22 03 77.
102		1974	to FAA 06 82	Mirage 5P2. Del. to FAP between 05 68 & 12 69; del. to FAA as C-603, 06 82.
103		1974	to FAA 06 82	Mirage 5P2. Del. to FAP between 05 68 & 12 69; del. to FAA as C-619, 06 82.
104		1974	to FAA 06 82	Mirage 5P2. Del. to FAP between 05 68 & 12 69; del to FAA as C-604, 06 82.
105		1974	to FAA 06 82	Mirage 5P2. Del. to FAP between 05 68 & 12 69; del. to FAA as C-607, 06 82.
106		1974	to FAA 06 82	Mirage 5P2. Del. to FAP between 05 68 & 12 69; del. to FAA as C-610, 06 82.
107		1974	to FAA 06 82	Mirage 5P2. Del. to FAP between 05 68 & 12 69; del. to FAA as C-628, 06 82.
108		1976	ret. 1997	Mirage 5P3. In storage.
109		1976	ret. 1997	Mirage 5P3. In storage at SEMAN, Las Palmas.
110		1976	ret. 1997	Mirage 5P3. In storage at SEMAN, Las Palmas.
111		1976	Unknown	Mirage 5P3
112		1976	Unknown	Mirage 5P3
113		1976	ret. 1997	Mirage 5P3. In storage at SEMAN, Las Palmas.
114		1976	Unknown	Mirage 5P3

Mirage 2000P/DP

Serial Number	c/n	Service Entry	Status	Remarks
050		1986	In service	Mirage 2000P. Blue/grey colour scheme
051	52	1986	In service	Mirage 2000P
052		1986	In service	Mirage 2000P
053		1986	In service	Mirage 2000P
054		1986	In service	Mirage 2000P
060		1986	In service	Mirage 2000P
061	126	1986	In service	Mirage 2000P
062	126	1986	In service	Mirage 2000P. Blue/grey colour scheme
063		1986	In service	Mirage 2000P
064		1986	In service	Mirage 2000P
193	43	1986	In service	Mirage 2000DP. Blue/grey colour scheme
195		1986	In service	Mirage 2000DP

7. VENEZUELA | Fuerza Aérea Venezolana (FAV, Venezuelan Air Force)

Mirage IIIEV

Serial Number	c/n	Service Entry	Status	Remarks
0240		1973	w/o 02 12 86	Crashed 02 12 86 due to fuel starvation, pilot ejected.
0624		1973	w/o 19 10 90	Crashed 19 10 90.
2473 (1)		1973	w/o	Crashed, date unknown. Replaced with new aircraft (Mirage 5EV?) with same serial number
2843		1973	w/o 27 10 90	Crashed 27 10 90.
3039		1973	w/o 05 08 86	Crashed 05 08 86, pilot ejected.
4058		1973		Upgraded to Mirage 50EV.
6732		1979		Upgraded to Mirage 50EV.
7712		1973	w/o 05 08 86	Crashed 05 08 86, fuel starvation, pilot ejected.
8940		1973	w/o 17 09 79	Crashed 17 09 79, pilot ejected.
9325		1973	w/o 05 08 86	Crashed 05 08 86, fuel starvation, pilot ejected.

Mirage 5V/DV

Serial Number	c/n	Service Entry	Status	Remarks
0155	086/470	1992		Mirage 5V. Upgraded to Mirage 50EV.
1225		1973	preserved	Mirage 5V. Retired in 1990. Preserved at El Libertador.
1297		1973		Mirage 5V. Upgraded to Mirage 50EV.
2473 (2)		1973		Mirage 5V. Upgraded to Mirage 50EV. Took part in coup attempt on 27 11 92.
5471		1973	w/o 14 05 90	Mirage 5DV. Crashed 14 05 90. Pilots ejected.
5706		1973		Mirage 5V; upgraded to Mirage 50DV.
7162		1973		Mirage 5V. Upgraded to Mirage 50EV.
7381		1973	w/o 22 09 76	Mirage 5DV. Crashed 22 09 76. Pilots ejected.
7512		1979		Mirage 5DV; upgraded to Mirage 50DV.
9510		1973	w/o 30 07 82	Mirage 5V. Crashed 30 07 82. Pilot ejected.

Mirage 50EV/DV

Serial Number	c/n	Service Entry	Status	Remarks
0155	086/470	1992	ret. 2009	Mirage 5V; upgraded to Mirage 50EV. Retired in 2009.
0160		1990	w/o 14 10 93	Mirage 50EV. Took part on the actions of 27 11 92. Crashed 14 10 93. Pilot ejected.
1297	174/102/647	1973	To FAE 10 09	Mirage 5V; upgraded to Mirage 50EV. To Ecuador 26 10 09.
2056		1992	w/o 06 04 93	Mirage 50EV. Crashed 06 04 93, pilot ejected.
2191		1993	ret. 2009	Mirage 50EV. Retired in 2009.
2212		1993	ret. 2009	Mirage 50EV. Retired in 2009.
2353		1993	To FAE 12 09	Mirage 50EV. To Ecuador 13.12.09.
2473		1973	ret. 2009	Mirage III or 5; upgraded to Mirage 50EV. Took part in coup attempt on 27 11 92.
3033		1993	ret. 2009	Mirage 50EV. Retired in 2009.
3373		1993	To FAE 12 09	Mirage 50EV. To Ecuador 13.12.09.
4058		1973	ret. 2009	Mirage IIIEV; upgraded to Mirage 50EV. Retired in 2009.
4212		1991	To FAE 10 09	Mirage 50DV. Colour scheme for 30th anniversary of Mirage in Venezuela. To Ecuador 26 10 09.
5145		1993	w/o 21 06 07	Mirage 50EV. Crashed 21 06 07. Pilot Capitán Aldao killed.
5706		1973	To FAE 12 09	Mirage 5V; upgraded to Mirage 50DV. To Ecuador 13.12.09.
6732	200	1979	w/o 17 09 04	Mirage IIIEV upgraded to Mirage 50EV. Crashed 17 09 04 at El Libertador, pilot ejected.
7162		1973	ret. 2009	Mirage 5V; upgraded to Mirage 50EV. Retired in 2009.
7512		1979	To FAE 10 09	Mirage 5DV; upgraded to Mirage 50DV. To Ecuador 26 10 09.
????			ret. 2009	Mirage 50EV.
????			ret. 2009	Mirage 50DV.

APPENDIX II

Camouflage patterns and primary armament of Latin American Mirages

1. ARGENTINA | Fuerza Aérea Argentina (FAA, Argentine Air Force)

The FAA Mirage IIIEA received the standard South-East Asia camouflage pattern as applied on USAF F-4s of the late 1960s. This consisted of tan (FS30219), dark green (FS34079) and green (FS34102) on the upper surfaces, with grey (FS36622) lower surfaces. Standard armament comprised Matra R.530F-1 and R.550 Magic 1 air-to-air missiles.

The latest camouflage pattern worn by FAA Mirages is relatively simple, consisting of an overall grey very similar to light ghost grey (FS36375). All markings are toned down and serial numbers are applied only on the front undercarriage bay doors. Only Magic air-to-air missiles remain in service.

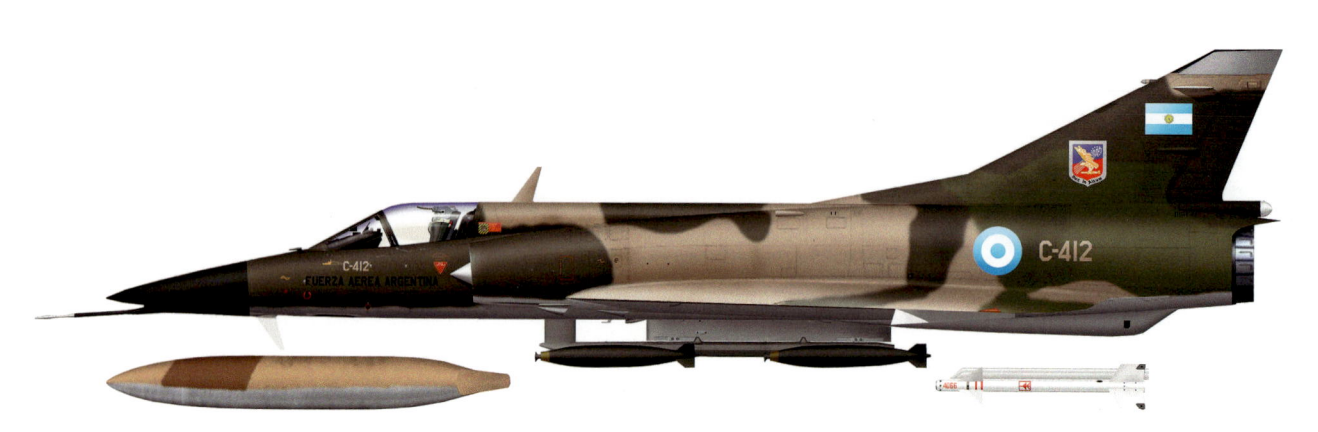

Daggers acquired from Israel in the late 1970s and early 1980s received a camouflage pattern very similar to that of the FAA Mirage IIIEA. This consisted of tan (FS30219), dark green (FS34079) and green (FS34102) on the upper surfaces, with grey (FS36622) on the lower surfaces. Standard armament comprised Israeli-made Shafrir II missiles (right lower corner) and US-made Mk 82 bombs, often carried on multiple-ejector racks under the centreline, as shown here.

Other weapons configurations during the Malvinas/Falklands War included the carriage of only two Mk 82s under the centreline pylon, or two additional Mk 82s on pylons on the rear undersides of the fuselage (sometimes, only the latter two were carried, since the centreline pylon was required for the carriage of additional drop tanks).

Surviving Daggers are now painted in a similar fashion to the Mirage IIIEA. In the last 20 years, the FAA has purchased a number of newer French GP bombs, including FAS 250 weapons (manufactured under licence in Argentina), up to two of which can be carried on the centreline. Dagger/Finger C418 is shown in its latest livery, including two kill markings for Royal Navy warships from the Malvinas/Falklands War.

C-404 was the second ex-Peruvian Mirage 5P3 to enter FAA service in June 1982 (the aircraft previously wore the FAP serial number 104). The aircraft initially retained its Peruvian camouflage colours, and was usually armed with R.550 Magic air-to-air missiles.

Remaining Mirage 5M Maras are now painted in the same camouflage pattern as FAA Mirages and Fingers. Of interest is their new dielectric radome, as well as aerials that resemble those on the Mirage IIIEA. In addition to locally manufactured GP bombs of French design, they are usually armed with Magic air-to-air missiles.

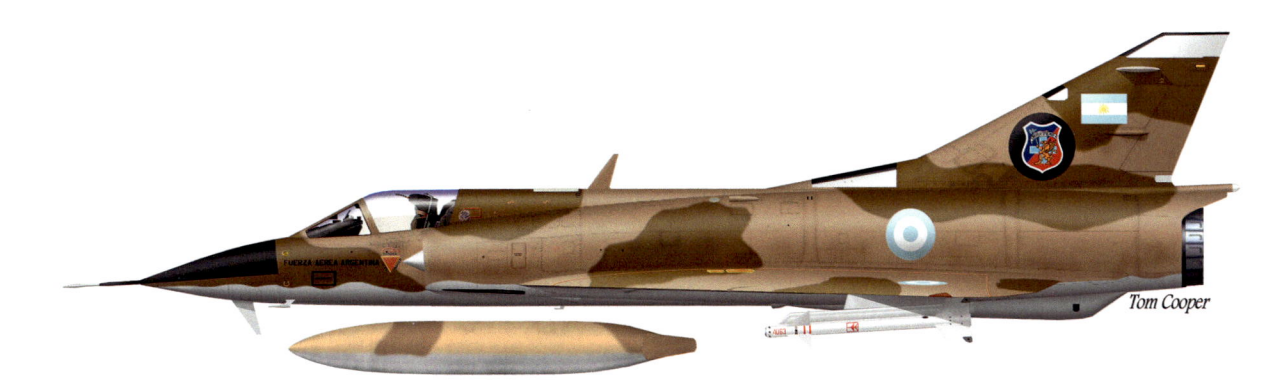

Immediately after delivery, the FAA Mirage IIIC received this unusual camouflage pattern consisting of tan (FS30219), plus a shade of tan (FS30051) with a strong green component (this usually rapidly deteriorated into green). Standard armament comprised Shafrir II missiles: with their Ibis radars replaced by a rangefinder only, these aircraft could not deploy R.530s.

2. BRAZIL | Força Aérea Brasiliera (FAB, Brazil Air Force)

Brazil's Mirage IIIE aircraft were originally painted in aluminium overall, and had large red markings around the engine intakes. Instead of a fin flash, national colours were applied on the rudder.

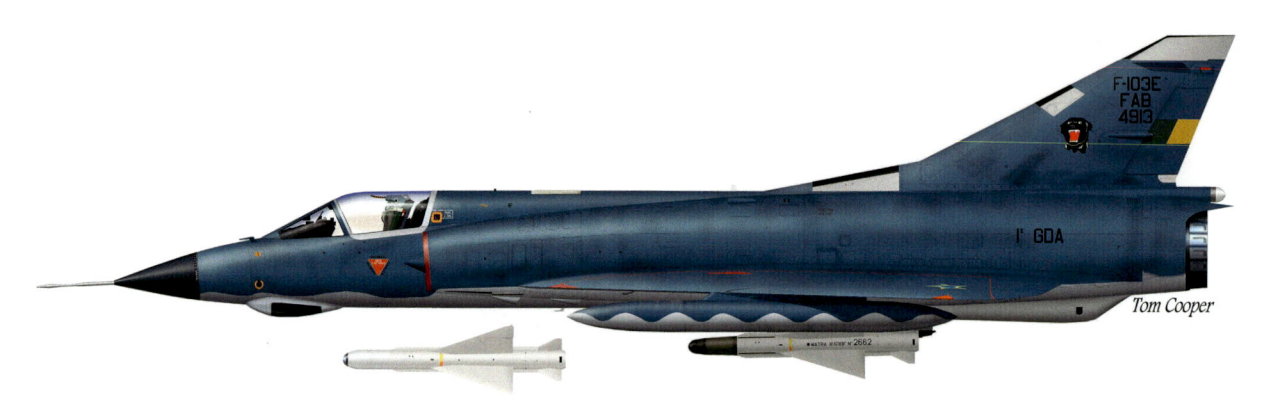

Later in its career with the FAB the Mirage IIIEBR received a camouflage scheme of gris bleu foncé (dark blue grey, similar to FS35164 or RAF7012) on the upper surfaces, and gris bleu clair (light blue grey) on the lower surfaces. R.530 missiles in their IR-homing (left) and radar-homing variants (centreline) represented the main armament in the 1960s and 1970s.

Later in their careers, Brazilian Mirages received a coat of darker blue grey (possibly gris bleu foncé II) on all upper surfaces, with light blue grey lower surfaces. The Israeli-made Python III replaced older French missiles.

Brazilian F-2000Cs (the local designation) wear the standard French Air Force camouflage pattern consisting of Celomer 1625 (gris/bleu moyen clair, or medium blue grey) and Celomer 1620 (gris/bleu-vert moyen foncé, or dark medium blue grey) on the upper surfaces, with Celomer 1625 on the undersides. Standard missile armament includes Matra R.550 Magic 2 and Super 530D air-to-air missiles.

3. CHILE | Fuerza Aérea de Chile (FACh, Chile Air Force)

Chilean Mirage 50FCs wore a standardised camouflage pattern consisting of gris vert foncé (bronze green) and gris bleu très foncé (dark grey) on the upper surfaces, with gris lucide (light grey, similar to FS26622) on the undersides. A similar scheme was adopted for early Mirage 5s delivered to Colombia and Peru.

Soon after their upgrade to Pantera standard, FACh Mirage 50s were painted in this camouflage pattern based on that of the Mirage 5s, though including colours very similar to those applied on more modern Mirage 2000s. Colours consisted of Celomer 1625 and Celomer 1620 on the upper surfaces, with light grey lower surfaces. Except for various French- and US-made weapons (including the Mk 84 GP bomb, right corner), they could be armed with up to four Python IIIs, Israeli-made Griffin LGBs and indigenous Cardoen CBU250s (left corner) and CBU500s. They were also equipped with Israeli-made Elta EL/L-8212 ECM pods (bottom centre).

While serving with the FACh, ex-Belgian Mirage 5BA Elkans retained the camouflage pattern applied before delivery. This consisted of tan (FS20219), green (FS24064) and light green (FS24102) on the upper surfaces, with light grey (FS26622) undersides.

4. COLOMBIA | Fuerza Aérea Colombiana (FAC, Colombian Air Force)

As originally delivered to Colombia, Mirage 5s were painted in a standardised camouflage pattern consisting of gris vert foncé (bronze green) and gris bleu très foncé (dark grey) on the upper surfaces, with gris lucide (light grey, similar to FS26622) on the undersides.

Following their upgrade with Israeli avionics, FAC Kfirs received a camouflage pattern apparently consisting of a colour very similar to aircraft interior grey (FS36231) and aircraft green (FS34031). This was applied on the upper surfaces according to the standardised French camouflage pattern for the Mirage III/5 series. Israeli-made Griffin LGBs entered service in addition to US-made Mk 82 GP bombs.

Israeli-made Kfirs wore a very similar camouflage pattern to that of FAC Mirage 5s, though in slightly darker colours. In addition to various French- and US-made GP bombs, Python III missiles were part of the Kfir's arsenal from the beginning of its service with the FAC.

229

In recent years, most surviving Colombian Kfirs have worn this simple camouflage pattern, consisting of dark grey blue (FS35237) on the upper surfaces and either light grey (either FS26622 or 36495) or light compass ghost grey (FS36375) on the lower surfaces.

The first of several Kfirs as painted while in the process of being upgraded for Colombia in Israel during 2009-10. The aircraft received a camouflage pattern apparently consisting of intermediate blue (FS35164) on the upper surfaces, with insignia white (FS17925) on the undersides. These aircraft will be made compatible with advanced Israeli stores including the Lizard LGB (left bottom) and Litening targeting pod (left centre).

On their delivery to Colombia in early 2010 most of the aircraft belonging to the latest batch of Kfirs were painted as shown here, in dark grey on the upper surfaces, and light grey on the lower surfaces. Notable is the radar nose and the slight 'bulge' in the upper fuselage, very similar to that of the South African Cheetah, the large numbers of warning markings around the fuselage and on the upper surfaces of both wings, and the fact that these aircraft are still equipped with Python III air-to-air missiles.

5. ECUADOR | Fuerza Aérea Ecuatoriana (FAE, Ecuadorian Air Force)

Ecuadorean Mirage F.1JAs were painted in this unique camouflage pattern, consisting of kaki (khaki green) and a green shade similar to vert pomme (light or apple green) on the upper surfaces, this being applied according to a standardised scheme. The lower surfaces appear to have originally been painted in aluminium grey, and later in light grey (FS26622). Except for French-made Magic 1s, they could carry a wide range of GP bombs (including US-made M117s), and were later made compatible with Python III and IV air-to-air missiles.

Ecuadorian Kfirs were originally painted in an unusual, disruptive camouflage pattern consisting of light green (FS34434) and dark green on the upper surfaces, with light grey (FS26622) undersides. They were armed almost exclusively with US-made GP bombs, such as the Mk 82 (including Mk 82s equipped with Mk 15 Snakeye retarding fins), and Israeli-made Python III and Shafrir II missiles.

Ecuadorian Kfir C10s appear to be camouflaged in ghost grey (FS36320) and light compass ghost grey (FS36375), similar to USAF F-15s of the 1980s, but some photographs indicate the use of light grey and compass ghost grey instead (as shown here). The aircraft are now regularly armed with Python IIIs and Python IVs (shown here).

231

The first photographs of the ex-Venezuelan Mirage 50EV in FAE service indicate that the aircraft retain the camouflage worn prior to delivery (see below for details). Obviously, an Ecuadorian fin flash replaces the FAV insignia.

Delivery of Cheetah Cs from South Africa to Ecuador had not taken place as this book was being prepared. However, the authors expect the aircraft to either remain in the same camouflage pattern as applied during their service with the SAAF, consisting of ghost grey (FS36320) and light compass ghost grey (FS36375), as depicted here, or to receive a camouflage similar to that of FAE Kfir CEs. They are likely to be armed not only with Python IIIs, but also Python IVs and Griffin LGBs.

6. PERU | Fuerza Aérea del Perú (FAP, Peruvian Air Force)

As originally delivered to Peru, Mirage 5s were painted in a standardised camouflage pattern consisting of gris vert foncé (bronze green) and gris bleu très foncé (dark grey) on the upper surfaces, with gris lucide (light grey, similar to FS26622) on the undersides.

Mirage 5P3 FAP 102 served for some 13 years with FAP before it was delivered to the FAA in June 1982. There it recieved the new serial number C-603.

Most FAP Mirage 5P3s and Mirage 5P4s received this camouflage pattern consisting of brun café (sand) and brun noisette (dark red-brown, which tended to bleach quite rapidly under the desert sun), applied according to a standardised scheme on the upper surfaces, with gris lucide (light grey, similar to FS26622) on the undersides.

233

Camouflaged in the same colours as the FAP Mirage 5P3, this Mirage 5P4 is shown together with French-made SAMP BL70 400kg (under the centreline) and SAMP Type 21C 400kg bombs (right corner). The artwork above (showing Mirage 5P3 FAP 106) also provides details of the Alkan drop tank/twin-tandem adapter, stressed for carriage of up to four 250kg bombs.

In addition to the refuelling probe, this Peruvian Mirage 5P4M shows the unique, French-made combination of a Matra rocket pod and drop tank, designated JL100 (left lower corner), the Matra RLF2 launcher for 6 68mm rockets (bottom centre) and the Matra 155 launcher for 18 68mm rockets (right lower corner). Thomson-Brandt LR 100-4 and 100-6 launchers, carrying four and six 100mm rockets, respectively, were also observed in service.

Painted in a similar fashion to most other Peruvian Mirages (except for its large, 'fake radome'), this Mirage 5P4M is shown with a Soviet-made R-3S air-to-air missile (AA-2 Atoll, left lower corner), and a French-made Magic 1 (right lower corner).

234

Peruvian Mirage 2000s were delivered wearing a camouflage pattern consisting of light sand and terre (chestnut brown) or brun noisette (dark red-brown), applied according to a standardised scheme on the upper surfaces, with Celomer 1625 on the undersides. Either US-made GBU-12 or Israeli-made Lizard LGBs entered service in more recent years (left lower corner).

More recently, following an overhaul, FAP Mirage 2000s have been re-painted in a camouflage pattern resembling that applied to Brazilian and French aircraft of this type. Although deliveries of Super 530D air-to-air missiles have been reported, primary armament remains the Magic 2 air-to-air missile.

7. VENEZUELA | Fuerza Aérea Venezolana (FAV, Venezuelan Air Force)

The FAV Mirage IIIEV received this very unusual camouflage pattern. Although based on the standardised scheme applied to most other exported Mirages, and consisting of the same colours as used on the Argentinean Mirage IIIEA – tan (FS30219), dark green (FS34079) and green (FS34102) on the upper surfaces, with grey (FS36622) lower surfaces – the two green shades were applied in a very unusual fashion. The aircraft were armed with R.530F-1 and US-made AIM-9E Sidewinder air-to-air missiles.

The FAV Mirage 5V received the same camouflage pattern as the Mirage IIIEV, though with softer borders. Weapons shown below the aircraft include the Alkan drop tank/twin-tandem adapter (left lower corner), which can carry US-made Mk 82s (shown here), various French weapons, and even GBU-12 LGBs. Shown bottom centre is a Durandal anti-runway bomb, and in the lower right corner is a Matra RLF2 launcher for six 68mm rockets.

Equipped with appropriate avionics, the FAV Mirage 50EV was the only Mirage III/5/50 variant compatible with the sophisticated Aérospatiale AM.39 Exocet anti-ship missile (one round is shown mounted on it special adaptor, under the aircraft centreline). However, the aircraft's only air-to-air missile was the Magic 1.

APPENDIX III

Primary weapons used by Latin American Mirages

	Guns		
DEFA 552	30mm		Mirage III, Mirage 5, Mirage 50, Dagger
DEFA 553	30mm		Kfir, Mirage F.1
DEFA 554	30mm		Mirage 2000

	Air-to-air missiles	
Matra R.530	Infra-red and semi-active radar homing (SARH), range 20km (12 miles), speed Mach 2.7	Mirage IIIE
Matra R.550 Magic 1 and 2	Infra-red, range 15km (9 miles), speed Mach 3	Mirage IIIE/D, Mirage 5, Mirage F.1, Mirage 2000, Mirage IIIC
Matra Super 550D	SARH, range 40km (24.9 miles), speed Mach 4.5	Mirage 2000
Vympel K-13 (AA-2 'Atoll')	Infra-red, range 8km (5 miles), speed Mach 2.5	Mirage 5P
Rafael Shafrir II	Infra-red, range 5km (3 miles), speed Mach 2.5	Mirage IIIB/C, Dagger/Finger, Mirage 50, Kfir
Rafael Python III	Infra-red, range 15km (9 miles), speed Mach 3.5	Mirage IIIE/D, Mirage 50, Kfir, Mirage F.1, Mirage 5COAM
Rafael Python IV	Infra-red, range 15km (9 miles), speed Mach 3.5	Kfir CE
AIM-9B Sidewinder	Infra-red, range 4.8km (3 miles), speed Mach 1.7	Mirage 50, Mirage IIIEBR, Mirage IIIEV
AIM-9P Sidewinder	Infra-red, range 18km (11 miles), speed Mach 2.5	Mirage 5M
AIM-9L Sidewinder	Infra-red, range 18km (11 miles), speed Mach 2.5	Mirage IIIEA
AIM-9M Sidewinder	Infra-red, range 18km (11 miles), speed Mach 2.5	Mirage IIIEA
Mectron MAA-1 Piranha	Infra-red, range 7km (4.5 miles), speed Mach 3.5	Mirage IIIEBR

	Air-to-ground missiles	
AS.20	Optically guided, range 10km (6 miles), speed Mach 1.7	Mirage 5P
AS.30	semi-active radio controlled range 12km (7 miles), speed Mach 1.5	Mirage 5P
Aérospatiale AM.39 Exocet	INS with ARH in terminal phase of flight, range 40 to 70km (25 to 45 miles), speed Mach 0.93	Mirage 5, Mirage 50

	Rockets	
70mm	FFAR	Mirage III, Mirage 5, Mirage 50, Mirage 2000, Mirage F.1, Dagger/Finger, Kfir

	Laser-guided bombs	
Elbit Lizard	500lb (227kg)	Mirage 5P, Mirage 2000P, Kfir
IAI Griffin	250, 500 and 1,000lb (113, 227 and 454kg)	Mirage 50 Pantera, Mirage 5COAM, Kfir
SAMP EU2	500lb (227kg)	Mirage 2000P

	Cluster bombs	
Cardoen CB 125-K	125kg (276lb)	Mirage 50
Cardoen CB 250-K	250kg (551lb)	Mirage 50, Mirage 5COAM, Kfir
Cardoen CB 500-K	500kg (1.102lb)	Mirage 50
FAS 300A and B	500lb (227kg) cluster bomb with 220 or 88 submunitions	Dagger/Finger, Mara, Mirage IIIC
	Anti-runway bombs	
Brant BAP.100	Parachute retarded and then boosted by a rocket for impact at about 260m/s. Eighteen 32kg (71lb) bombs providing a total weight of 710kg (1,565lb)	Mirage 2000P
Matra Durandal	220kg (483lb) rocket-boosted bomb, penetration at 260m/s	Mirage F.1
FAS 260	Based on BAP-100. Parachute retarded and then boosted by a rocket for impact at about 260m/s. Each bomb weighs 37kg (82lb)	Mirage IIIC, Mirage 5A, Dagger/Finger
	Specialised bombs	
FAS 850 Dardo 1	500lb (227kg) rocket-boosted standoff bomb, range 15km (9 miles)	Dagger/Finger, Mirage IIIC, Mirage 5A
FAS 850 Dardo 2	800lb (363kg) GPS-guided gliding standoff bomb, range 60km (37 miles), speed Mach 0.9	Dagger/Finger, Mirage IIIC, Mirage 5A
Brant BAT-120	Anti-armour weapon, based on BAP-100. Parachute retarded and then rocket-boosted for impact at about 260m/s. Eighteen 34kg (75lb) bombs providing a total weight of 73kg (161lb)	Mirage 2000P
FAS 280	Anti-armour weapon, based on FAS 260. Parachute retarded and then rocket-boosted for impact at about 260m/s. Each bomb weighs 34kg (75lb)	Dagger/Finger, Mirage IIIC, Mirage 5A
	Unguided bombs	
Explosivos Alaveses (Expal) BR-250	250kg (551lb); BRP version is parachute retarded	Mirage IIIEA, Mirage 5, Mirage IIIC, IAI Dagger/Finger
Explosivos Alaveses (Expal) BRI-400	400kg (882lb); BRIP is parachute retarded	Mirage 50
French 400kg HE	400kg (882lb)	Mirage 5P, Mirage F.1, Mirage 2000P
French 200kg HE	200kg (441lb)	Mirage 5P, Mirage F.1, Mirage 2000P
Argentine built	250lb (113kg)	IAI Dagger/Finger
Mk 81	250lb (113kg), with free-fall or Snakeye retarded tails	Mirage III, Mirage 5, Mirage 50, Mirage F.1, Mirage 2000, Kfir,
Mk 82	500lb (227kg), free-fall or Snakeye retarded tails	Mirage III, Mirage 5, Mirage 50, Mirage F.1, Mirage 2000, Kfir,
Mk 83	1,000lb (454kg), with free-fall or Snakeye retarded tails	Mirage III, Mirage 5, Mirage 50, Mirage F.1, Mirage 2000, Kfir,
French BL-9	250lb (113kg)	Mirage 5, Mirage 50, Mirage F.1
French BL-7	230lb (104kg)	Mirage 5, Mirage 50, Mirage F.1
Cardoen CFB-27	540lb (245kg)	Mirage 50
FAS 250	500lb (227kg), parachute retarded	Dagger/Finger, Mirage 5A, Mirage IIIC
FAS 800A	500lb (227kg) fragmentation bomb; optional parachute retarded tail	Dagger/Finger, Mirage 5A, Mirage IIIC
FAS 800B	500lb (227kg) fragmentation bomb; optional parachute retarded tail	Dagger/Finger, Mirage 5A, Mirage IIIC

Note: All FAS bombs were locally developed and built by the Argentine Air Force's Dirección de Sistemas. Some other countries used locally built bombs, these typically being based on US or European designs.

APPENDIX IV

Latin American Mirage units

1. ARGENTINA | Fuerza Aérea Argentina (FAA, Argentine Air Force)

IV Brigada Aérea

Grupo 4 de Caza
Mendoza, Mendoza province

Escuadrón 55
Mirage IIIC   1985 – 1991

VI Brigada Aérea

Grupo Aéro 6 de Caza
Tandil, Buenos Aires
province

Escuadrón 1
Finger and Dagger   1979 –

Escuadrón 2
Dagger   1981 – 1988
Mirage IIIEA 1988 –

Escuadrón 3
Mirage IIIDA/Dagger
two-seater/Mara) 1981 –

Escuadrón II Aeromóvil
'La Marinete'
San Julián
Dagger 25 04 – 25 08 82

Escuadrón III Aeromóvil
'Las Avutardas Salvajes'
Base Aeronavaloder
Almirante Quijada
Dagger 07 04 – 20 06 82

239

VIII Brigada Aérea

Grupo 8 de Caza
José C. Paz, Buenos Aires province

Escuadrón 1
Mirage IIIEA 1975 – 1988

Escuadrón 2
Mirage IIIEA 1978 – 1988

X Brigada Aérea

Grupo 10 de Caza
Río Gallegos, Santa Cruz province

Escuadrón Cruz y Fierro
Mirage III   1984 – 1986
Mirage 5   1986 – 1997

Comando de Material, Centro de Ensayos de Armamentos y Sistemas Operativos (CEASO)
Río Cuarto, Córdoba province
1991 – 2002

2. BRAZIL | Força Aérea Brasiliera (FAB, Brazil Air Force)

1° Ala de Defesa Aérea (1° ALADA)
Anápolis
Mirage III   1972 – 1979

1º Grupo de Defesa Aérea (1°GDA) 'Jaguar'
Anápolis
Mirage III   1979 – 2005

3° Força Aérea

1º Grupo de Defesa Aérea (1°GDA) 'Jaguar'
Anápolis
Mirage 2000   2006 –

3. CHILE | Fuerza Aérea de Chile (FACh, Chile Air Force)

IV Brigada Aérea, Grupo de Aviación N°4
Mirage 50
Base Aérea Los Cerrillos, Santiago   1980 – 1986
Base Aérea Chabunco, Punta Arenas   1986 – 2007

V Brigada Aérea, Grupo de Aviación N°8
Mirage 5M Elkan
Base Aérea Cerro Morena, Antofagasta   1995 – 2006

4. COLOMBIA | Fuerza Aérea Colombiana (FAC, Colombian Air Force)

Comando Aéreo de Combate No. 1
'CT. Germán Olano Moreno'
(CACOM-1)
BAM German Olan, Palanquero

Grupo de Combate 11

Escuadrón 212 'Mirage'
Mirage 5   1972 – ????

Escuadrón112 'Mirage'
Mirage 5   ???? –

Escuadrón 213 'Dardos'
Kfir   1988 – ????

Escuadron 111 'Dardos'
Kfir   ???? –

5. ECUADOR | Fuerza Aérea Ecuatoriana (FAE, Ecuadorian Air Force)

Ala de Combate 21
Base Aérea Taura, Guayaquil

Escuadrón de Combate 2112
'Cobras'
Mirage F.1   1979 –
Mirage 50   2009 –
Cheetah C   2011 –

Escuadrón de Combate 2113
'Leones'
Kfir   1982 –

242

6. PERU | Fuerza Aérea del Perú (FAP, Peruvian Air Force)

Grupo Aéreo No. 6
Base Aérea Capitán Quiñones
Gonzales, Chiclayo

Escuadrón de Caza 611
'Los Gallos'
Mirage 5 1968 – 2002

Escuadrón de Caza 612
'???'
Mirage 5 1968 – 1982

Grupo Aéreo No. 4
Base Aérea Coronel FAP Víctor
Maldonado Begazo, La Joya

Escuadrón de Caza-
bombardeo 411 '???'
Mirage 5 1997 – 2003

Escuadrón de Caza-
bombardeo 412 'Halcones'
Mirage 2000 1987 –

7. VENEZUELA | Fuerza Aérea Venezolana (FAV, Venezuelan Air Force)

Grupo Aéreo de Caza 11 'Diablos'
Base Aérea El Libertador, Palo Negro

Escuadrón de Caza 33
'Halcones'
Mirage III 1973 – 1991
Mirage 50 1991 – 2009

Escuadrón de Caza 34
'Caciques'
Mirage 5 1972 – 1991
Mirage 50 1991 – 2009

243

Appendix V

APPENDIX V

Latin American Mirage family

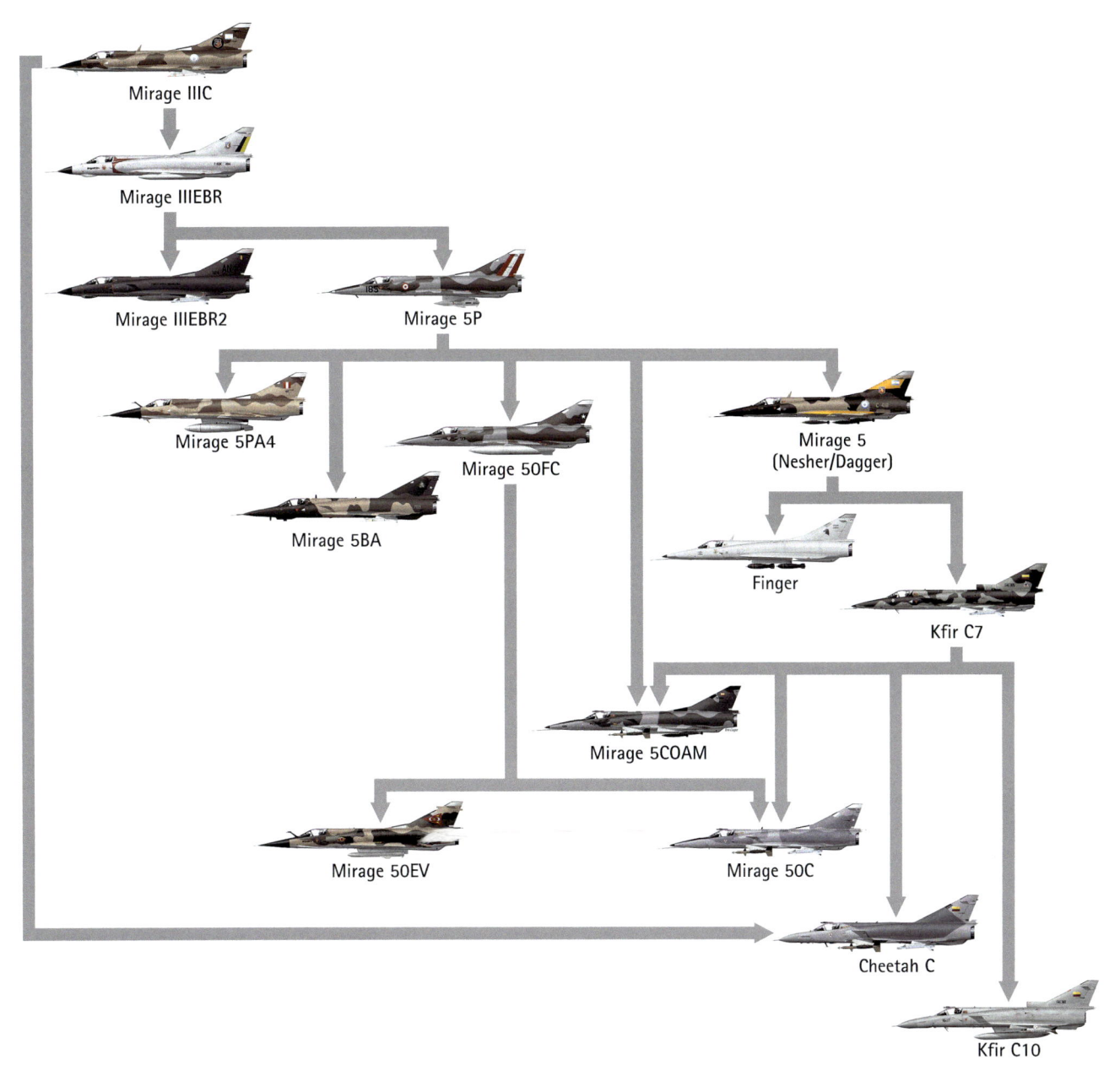

- Mirage IIIC
- Mirage IIIEBR
- Mirage IIIEBR2
- Mirage 5P
- Mirage 5PA4
- Mirage 50FC
- Mirage 5BA
- Mirage 5 (Nesher/Dagger)
- Finger
- Kfir C7
- Mirage 5COAM
- Mirage 50EV
- Mirage 50C
- Cheetah C
- Kfir C10

245

APPENDIX VI

Kits and Decals

Compiled with kind help of Mario Martínez, the following list includes most of the currently available plastic kits than can be used for replicating Latin American Mirage fighters.

Table 1: Aircraft Kits

Model	1/72 Scale Brand	1/48 Scale Brand	1/32 Scale Brand
AMD Mirage F.1	Airfix, Esci, Hasegawa, Heller	Esci, Italeri	n/a
Dassault Mirage IIICJ	Airfix, AML, High Planes, PJ Productions	Eduard, Hobbyboss	n/a
Dassault Mirage IIIE/EA/F-103(EDR)	Heller, High Planes, PJ Productions, Revell	Esci, Italeri	Revell
Dassault Mirage IIIDEA	High Planes, PJ Productions	n/a	n/a
Dassault Mirage 5/Elkan/Pantera	Heller, High Planes, PJ Productions	Esci	n/a
Dassault Mirage 2000P/2000-5	Airfix, Heller, Italeri	Eduard, Heller, Italeri, Monogram	n/a
Dassault Mirage 2000D	Heller, Italeri	Eduard, Heller	n/a
IAI Nesher/Dagger	High Planes, PJ Productions	n/a	n/a
IAI Kfir	Hasegawa, Italeri	Esci	n/a

Table 2: Decals

Manufacturer	Reference	Content & Comentary
Scale 1:72		
Albatros Decals	72-001 (out of print)	The Southern Flying Tigers - Peruvian MiG-29, Mirage 2000, Su-22
Albatros Decals	72-002 (out of print)	The Caribean Falcons - Cuban MiG-15/17/21/23/29 and Peruvian M2000
Atzec Models	D72-003	Ecuadorian Air Force II - A-37B, Strikemaster Mk 89, Mirage F.1
Atzec Models	D72-004 (out of stock)	Ecuadorian Air Force I - Jaguar, Kfir C2
Atzec Models	D27-005	Peruvian Air Force I - MiG-29, Mirage 2000
Atzec Models	D72-006 (out of stock)	Peruvian Air Force II - A-37, F-86F, Mirage 5, Su-22M-3
Atzec Models	D72-007 (out of stock)	Latin Eagles I - Argentinian A-4, Chile Hunters, Colombian Kfir and Mirage
Atzec Models	D72-021	Amazonia Mirages - Brazil and Colombia
Zotz	ZTZSP-3	Roundels of the World Part 3 - Central America
Zotz	ZTZSP-4 (sold out)	Roundels of the World Part 4 - South America

Scale 1:48		
Albatros Decals	48-001 (out of print)	The Southern Flying Tigers - Peruvian MiG-29, Mirage 2000, Su-22
Albatros Decals	48-002 (out of print)	The Caribean Falcons - Cuban MiG-15/17/21/23/29 and Peruvian M2000
Atzec Models	D48-003	Ecuadorian Air Force II - A-37B, Strikemaster Mk 89, Mirage F.1
Atzec Models	D48-004 (out of stock)	Ecuadorian Air Force I - Jaguar, Kfir C2
Atzec Models	D48-005	Peruvian Air Force I - MiG-29, Mirage 2000
Atzec Models	D48-006 (out of stock)	Peruvian Air Force II - A-37, F-86F, Mirage 5, Su-22M-3
Atzec Models	D48-007 (out of stock)	Latin Eagles I - Argentinian A-4, Chile Hunters, Colombian Kfir and Mirage
Atzec Models	D48-021	Amazonia Mirages - Brazil and Colombia
Atzec Models	D48-022	Venezuelan Air Force
Zotz	ZTZSP-3	Roundels of the World Part 3 - Central America
Zotz	ZTZSP-4 (sold out)	Roundels of the World Part 4 - South America

Scale 1:32		
Zotz	ZTZSP-3	Roundels of the World Part 3 - Central America
Zotz	ZTZSP-4 (sold out)	Roundels of the World Part 4 - South America

INDEX

1º Ala de Defesa Aérea 100, 101, 102
1º Grupo de Aviação de Caça 99
1º Grupo de Defesa Aérea 102, 110
1º/14º Grupo de Aviação 99
1º/4º Grupo de Aviação 99
1st Lieutenant Alberto Maggi 22, 90
1st Lieutenant Almoño 34
1st Lieutenant Carlos Antonietti 63, 64, 90
1st Lieutenant Carlos Bellini 95
1st Lieutenant Carlos Perona 17, 19, 20, 21, 88, 89, 91, 92
1st Lieutenant Carlos Sellés 22, 88
1st Lieutenant César Román 40, 41, 42, 43, 44, 45, 46, 48, 51, 54, 55
1st Lieutenant Daniel Valente 39, 55, 56
1st Lieutenant Daniel Gálvez 89
1st Lieutenant Demierre 39, 51, 54, 55
1st Lieutenant (later Captain) Dellepiane 39, 45, 47, 51, 54
1st Lieutenant Gabari Zocco 39, 51, 63, 64
1st Lieutenant Horacio Bosich 22
1st Lieutenant Horacio Mir González 34, 38, 56, 58, 60, 62, 63, 64, 66
1st Lieutenant (later Captain) Janett 34, 38, 59, 62, 63, 64
1st Lieutenant Jorge Ratti 39, 40, 61, 63, 64
1st Lieutenant Jorge Senn 38, 40, 44, 48, 49, 50
1st Lieutenant Luna 38, 58, 60, 61
1st Lieutenant Marcelo Puig 20, 22, 23, 89
1st Lieutenant Mario Callejo 38, 40, 48, 51, 53, 54, 55, 56, 67, 89
1st Lieutenant Musso 34, 39, 51, 55,
1st Lieutenant Norberto Dimeglio 34, 38, 40, 41, 42, 43, 44, 45, 46, 51, 53, 54, 55, 56, 90, 92
1st Lieutenant Piuma Justo 39
1st Lieutenant Roberto Yebra 20, 90
801 Naval Air Squadron 19, 20, 58

I Brigada Aérea (Chile) 116, 132
II Brigada Aérea (Argentina) 91
IV Brigada Aérea (Argentina) 13, 30, 71, 89, 90, 93, 133
V Brigada Aérea (Argentina) 28, 36, 71, 133
VI Brigada Aérea (Argentina) 13, 27, 35, 37, 56, 66, 69, 71, 75, 78, 79, 83, 84, 89, 179
VII Brigada Aérea (Argentina) 13
VIII Brigada Aérea (Argentina) 16, 19, 35, 69
IX Brigada Aérea (Argentina) 19, 35, 37, 38
X Brigada Aérea (Argentina) 68, 80, 89, 91

AA-2 Atoll 179
Able Seaman (M) Neil Wilkinson 62
Aero Commander AC-500U 36
Aerodrome Dr Mariano Moreno 14

Aeropuerto Arturo Merino Benítez 116
Aerospatiale AM.39 Exocet 48, 56, 198
Aérospatiale AS.330 Super Puma 199
Aérospatiale SA.315B Lama 90, 93, 133
Aérospatiale SA.330 Puma 60
AIM-9 Sidewinder 18, 21, 30, 50, 57, 58, 60, 61, 108, 115, 193, 194
Air Force General Francisco Visconti Osorio 199
AGM-45 Shrike 23
Ala de Combate 21 151, 155
Alferez Germán Demer 90, 92
Alferez Gustavo Rodríguez 90, 92
Alferez José Luis Correa 90, 92
Alferez Jorge Luis Valdez 90, 92
America States Organization 176, 177
Anápolis 99, 100, 102, 105, 109, 110, 111
Antofagasta 125, 132, 133
AN/APG-66 radar 29
AMX 71, 72, 106, 109, 124, 133
ARA San Luis 23
ARC Caldas 138
Área de Material Río IV 17, 27, 35, 37, 66, 70, 81, 82, 83, 84, 88, 89, 90, 91, 95, 96
Argentine Navy 17, 56
Armée de l'Air 31, 71, 100, 102, 104, 108, 109, 110, 111, 116, 117, 124, 174
ARV Almirante Brion 138
ARV General Urdaneta 138
ARV Mariscal Sucre 138
AS.30 55, 65, 73, 78, 173, 175, 179
Atlas Cheetah 124, 129, 139, 169, 170
Aviación Militar Bolivariana de Venezuela 203
Avro Vulcan 17, 19, 21, 23, 24, 104

BAe 125 42, 55
BAe Harrier GR.Mk 3 24
BAC Canberra 24, 38, 55, 87, 116, 151, 153, 155, 165, 173, 177, 178, 184, 193
BAC Strikemaster 151, 153, 160
BAM Comodoro Rivadavia 17
BAM Río Gallegos 17, 19, 20, 22, 24, 36, 37, 38, 45, 55, 56, 71, 80, 81, 82, 89, 91, 92, 93, 95
BAM San Julián 38, 40, 44, 45, 48, 51, 53, 55, 57, 59, 60, 61
Barquisimeto 193, 199, 200
Base Aérea Capitán Montes 175
Base Aérea Capitán Quiñónez González 34, 174
Base Aérea Capitán Renán Elías Olivera 184
Base Aérea Carlos Ibáñez 116, 119

Base Aérea Cerro Moreno 125, 132, 133
Base Aérea de Canoas 71, 109, 110, 124
Base Aérea de Manta 153
Base Aérea de Santa Maria 108
Base Aérea de Taura 151, 152, 153, 155, 156, 160, 162, 203
Base Aérea El Libertador 193, 199, 201, 202, 203
Base Aérea Francisco de Miranda 201
Base Aérea Madrid 140
Base Aérea Militar Germán Olano 137
Base Aérea La Joya (Coronel Maldonado Begazo) 176, 181, 184, 186, 189, 190
Base Aérea Las Palmas 174, 179, 181, 182, 185
Base Aérea Los Cóndores 72, 116, 125, 126,
Base Aérea Teniente Vicente Landaeta Gil 193, 199
Base Aeronaval Almirante Quijada 24
Base Oficial de Aviación Civil 14
Beagle Channel 16, 35, 115
Bell 212 133, 153, 177
Bell UH-1H 83, 110, 125, 199
Boeing 707 24, 25, 38, 51, 71, 106, 124, 125, 141
Brigadier General Grafigna 35
British Aerospace Sea Harrier FRS.Mk 1 17, 19, 20, 21, 22, 24, 38, 41, 44, 49, 50, 51, 52, 54, 57, 58, 60, 61, 62, 64

Captain Aldao 202
Capitán Alez Padilla 168
Captain Amílcar Cimatti 35, 38, 56, 58, 59, 60, 61, 62, 63
Captain Arnau 23, 24, 34
Captain Asenio 182
Captain Carlos Moreno 38, 56, 57, 60, 61, 62, 63, 64,
Captain (later Major) Carlos Napoleón Martínez 34, 37, 38, 56, 58, 59, 60, 61, 63, 64,
Captain Carlos Uzcategui 156, 157
Captain César Gonzalo Luzza 174, 175
Captain Donovan Bartolini Martínez 184
Captain Eduardo Carrera 153
Captain Eliades Moreno 141
Captain Fernando Robledo 68
Captain Gambarini 180
Captain Guido Moya 162
Captain Guido Zavalaga Ortigosa 184
Captain Guillermo Díaz Muñoz 138
Captain Guillermo Donadille 34, 44, 46, 48, 49, 50, 90
Captain Guillermo Ballesteros 22, 23, 24, 88
Captain González (FAA) 23, 24, 34,
Captain Gustavo Bucheli 151, 152
Captain Gustavo Cuesta 154
Captain Gustavo García Cuerva 19, 20, 21, 25, 42
Captain Hugo Acosta 141
Captain Jesús María Bevilaqua Paulosky 201
Captain Jorge Huck 22, 23
Captain Jorge Testa 17
Captain Jorge A. Suárez 141
Captain Juan C. Vélez 141
Captain Juan Fernando Correa Hernández 141
Captain Juan Manuel Grisales 145, 146
Captain Justet 69
Captain Kajihara 34, 89
Captain Labarca 199
Captain Luis López 151

Captain Maffeis 55, 56, 62, 63
Captain Marcos Czerwinski 19, 22
Captain Marco Estrella 151, 152
Captain Mario Pergolini 34
Captain Mauricio Mata 162, 163
Captain Miguel A. Barrera D. 141
Captain Norberto Cayetano Prior 66
Captain (later Major) Puga 34, 46, 48, 51, 52, 53,
Captain Patricio Velazco 168
Captain (later Commander) Patrnogic 180, 182
Captain Rafael Velosa A. 141
Captain Ram Brier 141
Captain Raúl Díaz 38, 46, 47, 48, 51, 52, 53
Captain Raúl Gambandé 17, 20
Captain Raúl Gómez 70
Captain Ricardo Vílchez Raa 184
Captain (later Commander) Rodríguez (FAP) 180, 182
Captain Rhode 37, 56, 58, 59, 60, 61, 62, 63, 64
Captain Wilson Salgado 154
Cardoen 119, 142
Cenepa 154, 155, 163, 180, 186
Centro Integrado de Defesa e Controle de Tráfego Aéreo 102, 103
Cessna A-37B Dragonfly 72, 115, 116, 125, 133, 151, 153, 155, 156, 158, 162, 177, 179
Chiclayo 34, 174, 177, 178, 187, 190
Chief Petty Officer Lionel N. Kurn 47
Chief Petty Officer David Heritier 47
CEASO 70, 95
Ceselsa 27
Colombian Navy 138
Colonel Jorge Luis Chaparro Pinto 190
Commander Fernando Hoyos 163
Commander Hilario Valladares 163
Commander Víctor Maldonado 157
Comando Aéreo de Combate 1 137, 141
Comando de Defesa Aérea 100
Comodoro Claudio Correa 90
Comodoro Manuel Mir 89
Comodoro Romeo Gallo 90
Coronel-Aviador Antônio Henrique Alves dos Santos 100
CRUZEX 71, 109, 110, 124
Cuba 102, 103, 194

Dassault Falcon 20 183, 184, 202
Dassault Super Etendard 38, 56, 58, 59
de Havilland Vampire 193
de Havilland Venom 193
Douglas A-4 Skyhawk/Fightinghawk 19, 21, 23, 27, 28, 29, 30, 31, 36, 38, 47, 49, 58, 59, 63, 67, 70, 71, 72, 73, 87, 110, 115, 116, 133, 140
Douglas C-47 36, 143

Ejército de Liberación Nacional 138
Ejército del Aire 29, 74
Elta EL/M-2001 ranging radar 36, 81, 119, 139, 140, 160,
Elta EL/M-2032 166, 170
Embraer 108, 109, 110
Embraer C-95 Bandeirante 110
Embraer EMB-312/T-27 Tucano 109, 110, 124, 143, 199, 201
Empresa Nacional de Aeronáutica (ENAER) 119, 120, 133

Escuadrón 212 137, 147
Escuadrón 2112 154, 156
Escuadrón 2113 154, 160, 162
Escuadrón 33 'Halcones' 193, 198, 202, 203
Escuadrón 34 'Caciques' 193, 198, 202
Escuadrón de Caza 11 160
Escuadrón de Caza 411 181, 185, 186
Escuadrón de Caza 412 184, 187, 190
Escuadrón de Caza 611 34, 174, 177, 180
Escuadrón de Caza 612 176, 177, 179, 181
Escuadrón de Electrónica 405 189
Escuadrón Mariano Moreno 14
Escuadrón I de Caza Interceptora 15
Escuadrón X Cruz y Fierro 71, 80, 84, 89, 90
Escuadrón 55 80, 90, 91, 92, 93, 95
Escuela de Aviación Militar 35
Estrecho de San Carlos/Falkland Sound 22, 23, 24, 42, 44, 46, 48, 51, 52, 53, 54, 56, 60, 61, 62, 63, 64,
Explosivos Alaveses 38, 56, 76, 187

Fábrica Militar de Aviones (FMA) 25
Fénix Squadron 22, 42, 55, 63
Feria Internacional del Aire y el Espacio (FIDAE) 186
Flight Lieutenant Bertie Penfold 58
Flight Lieutenant Paul Barton 20, 21
FMA IA-50 GII 90
FMA IA-58 Pucará 22, 23, 47, 67, 72, 73, 87, 133
FMA IA-63 Pampa 72, 93, 133, 186
FMA/Morane-Saulnier MS.760 Paris 67, 72, 90, 93, 133
Fokker F28 38
Fokker F27 133
Fuerza Aérea Sur 19, 48, 54
Fuerzas Armadas Revolucionarias de Colombia 138, 141, 143, 145

Gates Learjet 19, 22, 23, 61, 62, 63, 64, 91, 125, 176, 177, 178
General Dynamics F-16 Fighting Falcon 28, 29, 30, 70, 71, 72, 74, 75, 106, 108, 110, 125, 126, 129, 130, 133, 134, 138, 183, 190, 194, 199, 200, 201
General Jaime Estay Riveros 132
Gerald Resal 15
Gloster Meteor 13, 99, 115, 151
Gran Malvina/West Falkland 41, 44, 47, 48, 49, 51, 52, 53, 62
Grupo I de Vigilancia Aérea Escuela 13, 101
Grupo II de Vigilancia y Control del Espacio Aéreo 67, 83, 92
Grupo 4 de Caza 38, 90
Grupo 6 de Caza 35, 36, 37, 54
Grupo 8 de Caza 16, 17, 35
Grupo 10 de Caza 84, 89
Grupo Aéreo No. 4 (FAP) 184, 188, 190
Grupo Aéreo No. 6 (FAP) 177, 179
Grupo Aéreo No. 7 (FAP) 177
Grupo Aéreo No. 9 (FAP) 177, 184
Grupo Aéreo No. 11 (FAP) 177, 178, 179, 181, 189, 190
Grupo Aéreo Amazonas 155
Grupo de Aviación No .3 134
Grupo de Aviación No. 4 (Chile) 116
Grupo de Aviación No. 4 (Perú) 116, 185
Grupo de Aviación No. 8 132, 134
Grupo de Caza 212 151, 152, 153
Grupo de Caza No. 11 193

Grupo de Caza No. 12 193
Grupo de Combate 211 151
Guayaquil 151, 152

Hawker Hunter 115, 116, 117, 129, 133, 173, 174, 175,
Heyl Ha'Avir (IDF/AF) 14, 16, 33, 34, 36, 86, 139, 140,
HMS *Alacrity* 19, 41, 45
HMS *Antrim* 46, 47, 60
HMS *Ardent* 48, 61
HMS *Argonaut* 60
HMS *Arrow* 19, 41, 45
HMS *Brilliant* 48, 49, 59, 60, 61
HMS *Broadsword* 51, 52, 60
HMS *Cardiff* 24, 56
HMS *Coventry* 51, 52
HMS *Exeter* 64
HMS *Glamorgan* 19, 41, 43, 45, 60
HMS *Glasgow* 59
HMS *Hermes* 52
HMS *Intrepid* 62
HMS *Invincible* 50, 58
HMS *Plymouth* 64

IAI Nesher 16, 33, 34, 36, 139
IAI Griffin 142, 143, 166
IAI Lizard 180, 187
Israel 13, 14, 33, 34, 35, 36, 51, 65, 66, 75, 76, 86, 87, 88, 89, 90, 92, 108, 110, 116, 119, 139, 140, 141, 148, 151, 154, 160, 167, 170, 173, 180, 187
Israel Aircraft Industries 33, 34, 86, 119, 139, 140, 142, 148, 151, 160, 165, 166, 170, 180
Ilyushin Il-62 102, 103
Iquique 72, 116, 125
Isla Soledad/East Falkland 42, 62
Islas Malvinas/Falklands 17, 18, 19, 20, 21, 23, 25, 37, 41, 42, 44, 57, 78, 82, 90, 102, 117, 179

Jean Bongiraud 152
John Keith Watson 60

Le Bourget 184, 188
Lieutenant Juan Bernhardt 58, 59, 60, 61, 62, 63
Lieutenant Brian Haigh 58
Lieutenant Carlos Castillo 46, 51, 52, 53
Lieutenant Carlos Maroni 90, 92
Lieutenant David Smith 52
Lieutenant Gustavo Aguirre Faget 40, 41, 42, 43, 44, 46, 47, 51, 53, 54, 55, 56
Lieutenant Héctor Volponi 38, 56, 57, 59, 60, 61
Lieutenant (later Captain) Hernán Ayala 151, 152, 154
Lieutenant Joana Ximena Herrera 143
Lieutenant (later 1st Lieutenant) José Ardiles 37, 38, 58
Lieutenant (later Major) Luis Briatore 84, 90, 92
Lieutenant Martin Hale 58, 60, 61
Lieutenant Marco López 162
Lieutenant Marco Palacios 167
Lieutenant Pacheco Peña 201
Lieutenant Pedro Bean 56, 59, 60
Lieutenant Percy Juan Ryberg 90, 92
Lieutenant Raúl Estevez 90, 92

251

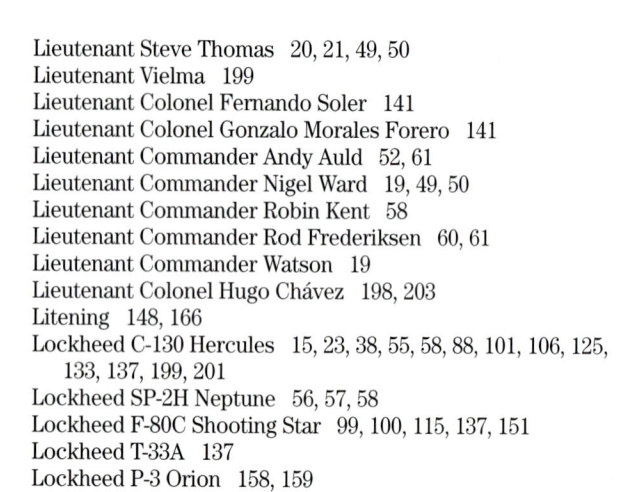

Lieutenant Steve Thomas 20, 21, 49, 50
Lieutenant Vielma 199
Lieutenant Colonel Fernando Soler 141
Lieutenant Colonel Gonzalo Morales Forero 141
Lieutenant Commander Andy Auld 52, 61
Lieutenant Commander Nigel Ward 19, 49, 50
Lieutenant Commander Robin Kent 58
Lieutenant Commander Rod Frederiksen 60, 61
Lieutenant Commander Watson 19
Lieutenant Colonel Hugo Chávez 198, 203
Litening 148, 166
Lockheed C-130 Hercules 15, 23, 38, 55, 58, 88, 101, 106, 125, 133, 137, 199, 201
Lockheed SP-2H Neptune 56, 57, 58
Lockheed F-80C Shooting Star 99, 100, 115, 137, 151
Lockheed T-33A 137
Lockheed P-3 Orion 158, 159

Major Augusto Romero-Lovo Ferrecio 174
Major Aviador Paulo César Pereira 102
Major Carlos Luna 24
Major Enrique Caballero Orrego 157
Major Flavio E. Ulloa E. 141
Major Felipe Conde Garay 184
Major Fernando Medrano J. 141
Major Guillermo Olaya 147
Major Gregorio Mendiola 163, 164
Major Héctor Heredia 151
Major Héctor Jimmy Mosca Sabate 186
Major Hernán Quiroz 160
Major Hernandez 152
Major Higinio Robles 38, 56, 58, 59, 60, 61, 62, 64, 89
Major Javier Gamboa 180
Major Juan C. Ramírez M. 141
Major José Sánchez 19, 22, 23, 24
Major José Orlando Bellon 103
Major Juan Sapolski 34, 37, 38, 40, 46, 54, 67, 88
Major Juan Vivero 162
Major Miguel Camacho M. 141
Major Patricio González 151
Major Páez 19
Major Piuma 46, 48, 49, 50
Major Puga 34, 46, 48, 51, 52, 53
Major Raúl Banderas 156, 157
Major Silva 21
Major Vincenzo Sicuso 84
Major Walter Milenko Vojvodic Vargas 186
Major William Birkett 153
Maracay 193
Matra R.530 16, 25, 34, 101, 115, 152, 193
Matra R.550 16, 25, 82, 105, 152, 179, 187, 194, 198
McDonnell Douglas F-4 Phantom II 13, 82, 99, 100, 139
McDonnell Douglas F/A-18 Hornet 28, 70, 189
McDonnell Douglas MD-80 200
Mectron MAA-1 Piranha 109
Mendoza 13, 30, 71, 80, 84, 85, 89, 90, 92, 93, 133
MiG-23 183, 194
MiG-29 108, 165, 181, 187, 189, 190
MirSIP 129, 130, 132
Mont-de-Marsan 174, 184, 186

North American F-86F Sabre 13, 27, 33, 67, 90, 93, 116, 137, 173, 193
North American F-100 Super Sabre 13, 33
Northrop F-5 Freedom Fighter/Tiger 13, 71, 72, 99, 100, 104, 106, 109, 110, 115, 120, 124, 125, 126, 193, 199, 200

Palanquero 37, 143, 146, 148
Puerto Argentino/Stanley 19, 21, 24, 40, 41, 42, 44, 56, 57, 64, 78, 87
Punta Arenas 82, 116, 119, 121, 126, 133

Qatar Air Force 29

RAF Mount Pleasant 82
Rafael Shafrir II 16, 34, 35, 38, 39, 40, 44, 55, 57, 58, 59, 65, 67, 76, 93, 115, 117, 121, 160, 161, 163, 165
Rafael Python III 108, 109, 121, 148, 158, 165, 170
Rafael Python IV 166, 167
Río Grande 24, 37, 38, 45, 46, 51, 56, 57, 58, 59, 60, 62, 63, 64
Rockwell OV-10 Bronco 143, 162, 199, 200, 201
Royal Jordanian Air Force 31, 75
Royal Navy 17, 19, 20, 56

Salitre 72, 125, 133
Sea Cat 61, 64
Sea Dart 24
Sea Slug 60
Sea Wolf 48, 60
Servicio de Mantenimiento (SEMAN) 175, 179, 181, 182, 184
SEPECAT Jaguar 129, 151, 153, 154, 155, 160, 169, 178
Sistema de Defesa Aérea e Controle de Tráfego Aéreo 100, 102
Six-Day War 13, 86, 87, 100, 116, 173
SNECMA Atar 9 33, 34, 89, 117, 132, 139, 170, 175, 198
South Africa 124, 129, 139, 169, 170
Sukhoi Su-22 115, 155, 156, 157, 158, 161, 162, 163
Sukhoi Su-25 165, 187
Sukhoi Su-27 129
Switzerland 33, 173

Tandil 27, 35, 36, 37, 38, 39, 40, 45, 51, 53, 55, 56, 58, 59, 63, 64, 65, 66, 68, 73, 78, 79, 84, 89, 92
Tenente-Aviador Roberto de Medeiros 102
Teniente Coronel Aviador Mauro Lazzarini de A. Silva 102
Thompson-CSF Cyrano Ibis 89
Thompson-CSF Cyrano II 16, 25, 26, 28, 31, 101, 193
Thompson-CSF Cyrano IV 116, 151, 176, 179, 198

United Kingdom (UK) 78, 84, 99, 102, 117, 179
USAF 73, 106, 133, 190
US Navy 139, 158, 159, 189

Venezuelan Navy 138, 198
Vicecomodoro Luis Villar 37, 46, 54, 55

Westinghouse AN/TPS-43 73, 83
Westland Lynx 42, 49, 56

Yom Kippur War 33, 34, 86, 87, 139

Zaire 198

ACIG.ORG

Online since 1999
ACIG.org is a multi-national project
dedicated to research about
air wars and air forces since 1945

Associated authors, photographers, artists and contributors
have published 16 books, dozens of articles and hundreds of artworks.
Multiple research projects are going on and we are
looking forward for your contributions:
join us at ACIG.org forum!

www.acig.org

AVIATIONGRAPHIC.COM
High Quality Profile Illustrations

AVIATIONGRAPHIC.com, guild of Aviation Artists & Illustrators has a wide on-line showcase. Our catalogue has TWO THOUSAND (2.000!) *Squadron Prints, Lithographs, Illustrations and Aviation Arts.* Thanks to the cooperation of reserachers, military technicians we create accurate original aircraft color artworks for publishing companies worldwide.

We proudly create the official lithos for U.S.Air Force, U.S.NAVY, Luftwaffe, KLU, HEER, AMI, TuAF, Brazilian AF and lots of Squadrons and Law Enforcement Units all over the World: for them we made and make the always rare Squadron Prints!

HARPIA PUBLISHING

Glide With Us Into The World of Aviation Literature

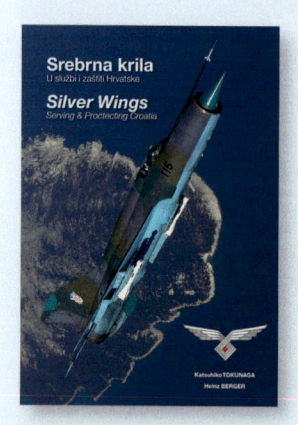

Silver Wings – Serving & Protecting Croatia
Katsuhiko Tokunaga and Heinz Berger
160 pages, 30x22 cm, hardcover with jacket
48.00 Euro   ISBN 978-0-9825539-1-6

The world-famous Japanese aviation photographer Katsuhiko Tokunaga covers the activities of today's Croatian Air Force in his well known and destinctive, nearly artistic style.

Following a brief introduction into the history of Croatian Air Force from 1991 until 2009, this exclusive, top quality photo monography provides dozens of high-quality photographs of the aircraft currently in service, and rich detail about the life, work and action of the men and women serving and protecting Croatia.

Appendices list the technical data of all aircraft in service and the current order of battle, together with all the unit insignia.

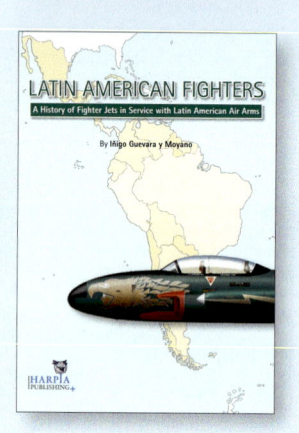

Latin American Fighters – A History of Fighter Jets in Service with Latin American Air Arms
Iñigo Guevara y Moyano
256 pages, 28x21 cm, softcover
35.95 Euro   ISBN 978-0-9825539-0-9

This book for the first time describes the military fighter jet aviation in Latin America. It covers the eventfull history of fighter jets in 17 countries ranging from Mexico in the north down to Argentina in the south. Each country is covered type by type in chronological order. Information on each type is being provided related to purchase, squadron service, losses, upgrades and service history. Each type ends with a table covering the number of delivered aircraft, different types and subtypes, delivery dates and known serial numbers. Each of the over 100 aircraft types mentioned could be covered with at least one picture.

An appendix lists the existing plastic scale model kits in 1/72, 1/48 and 1/32 scale as wells as decal sheets in regards to the 17 Latin American air forces featured in the book.

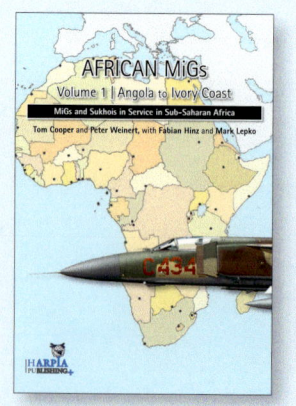

African MiGs Vol. 1 | Angola to Ivory Coast – MiGs and Sukhois in Service in Sub-Saharan Africa
Tom Cooper and Peter Weinert, with Fabian Hinz and Mark Lepko
256 pages, 28x21cm, softcover
35.95 Euro   ISBN 978-0-9825539-5-4

This second, expanded and fully revised edition of the groundbreaking book *African MiGs* examines the role and deployment history of MiG- and Sukhoi-designed fighters – as well as their Chinese derivatives – in no fewer than 23 air forces in Sub-Saharan Africa. This first volume, covering 12 air arms from Angola to Ivory Coast, will be followed by a second volume in 2011. In order to ensure precise documentation of every airframe delivered to and operated by the various air forces, special attention is given to illustrations as well as extensive tables of known serial numbers and attrition. The new volume is updated with much exclusive information, photographs and artworks. As such, it provides the most comprehensive and reliable source on the background of each of the features air forces, their organisation and unit designations, deliveries of fighters built by MiG, Sukhoi, Chengdu and Shenyang, camouflage, markings and combat deployment.

THE AVIATION BOOKS OF A DIFFERENT KIND
UNIQUE TOPICS I IN-DEPTH RESEARCH I RARE PICTURES I HIGH PRINTING QUALITY

www.harpia-publishing.com

HARPIA PUBLISHING

Glide With Us Into The World of Aviation Literature

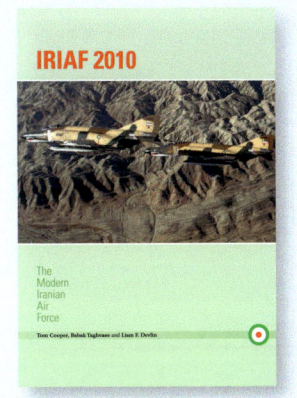

IRIAF 2010 – The Modern Iranian Air Force
Tom Cooper, Babak Taghvaee and Liam F. Devlin
160 pages, 30x22cm, softcover
29.95 Euro   ISBN 978-0-9825539-3-0

This richly illustrated book describes the current organisation and equipment of the Islamic Republic of Iran Air Force (IRIAF). Drawing on a wide range of digital photographs, IRIAF 2010 presents all types of aircraft currently operated by the IRIAF, many of which are supported by captions detailing individual aircraft histories. Following a summary of the air force's development since its early days in the 1920s, the centrepiece of this volume are 12 chapters that cover all major IRIAF bases and flying units stationed there, as well as a summary of the order of battle, in which all units are also represented in the form of patches worn by their pilots. Covering an often under-reported and misinterpreted topic, and one that is directly influential in the current standoff between Iran, the US and Israel, this book is a unique reference source for scholar and enthusiast alike.

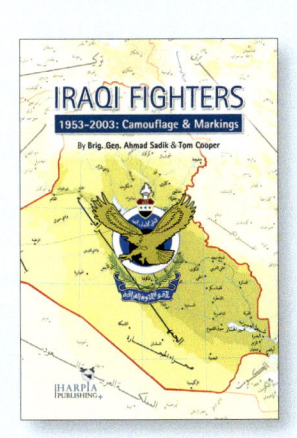

Iraqi Fighters 1953–2003 Camouflage & Markings
Brig. Gen. Ahmad Sadik and Tom Cooper
156 pages, 28x21cm, softcover
29.95 Euro   ISBN 978-0-615-21414-6

Richly illustrated with photographs and artworks, this book provides an exclusive insight into service history of 13 fighter jet types – from Vampires and Hunters to MiG-29s and Su-24s – that served with Royal Iraqi Air Force (RIrAF) and Iraqi Air Force (IrAF) between 1953 and 2003.
The result is a detailed history of RIrAF and IrAF markings, serial numbers and camouflage patterns, the in-depth history of each Iraqi fighter squadron, their equipment over the time as well as unit and various special insignias.
An appendix lists the exisiting plastic scale model kits in 1/72, 1/48 and 1/32 scale as well as decals sheets in regards to Iraqi Air Force.

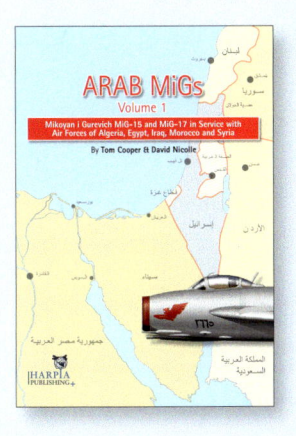

Arab MiGs Volume 1, MiG-15s and MiG-17s, 1955–1967
Tom Cooper and David Nicole
256 pages, 28x21 cm, softcover
35.95 Euro   ISBN 978-0-9825539-2-3

This study – the first in a series of similar publications – provides a unique and previously unavailable insight into the service of both types with five Arab air forces, including Algeria, Egypt, Iraq, Morocco and Syria. It tells the story of people that flew MIG-15s and MiG-17s, several of whom became dominant political figures in most recent history of these countries, and completes this with a review of combat operations in Yemen, as well as in three wars between the Arabs and the Israelis. Over 200 photos, colour artworks, maps and tables illustrate the story of the aircraft and their crews, as well as unit insignia in unprecedented detail. Extensive lists of serial- and construction numbers are provided as well.

THE AVIATION BOOKS OF A DIFFERENT KIND
UNIQUE TOPICS I IN-DEPTH RESEARCH I RARE PICTURES I HIGH PRINTING QUALITY

www.harpia-publishing.com

Dutch Aviation Society
P.O.Box 75545
1118 ZN Schiphol
The Netherlands
Fax: +31 (0) 84 - 738 3905
E-mail: info@scramble.nl
www.scramble.nl

The **Dutch Aviation Society** is a non-profit organisation run totally by volunteers. For those of you who have never heard of us, we will briefly explain our activities.

The main activities of the **Dutch Aviation Society** are:

- The publication of the monthly magazine **'Scramble'**.
- Maintaining the aviation website www.scramble.nl.
- To organise spotter conventions.
- Maintaining an aviation information database.
- Publishing from an aviation information database.

The production of the magazine, **Scramble**, is our core business. The magazine averages around 144 pages and more than 100 photographs from all over the world. It is published in the English language. It covers all aspects of civil and military aviation worldwide in many separate sections:

- Extensive civil airport and military airbase movements from the Netherlands;
- Civil and military movements from many European airports and airbases;
- Civil aviation news word wide (general news, jetliners, propliners, commuters, bizjets, bizprops, helicopters, extensive Soviet coverage, vintage aircraft, wrecks & relics);
- Dustpan & Brush (Stoffer & Blik), in depth reports about accidents and incidents worldwide;
- Military aviation news word wide (general news, procurement plans, unit changes, updates, orders of battle, vintage aircraft, wrecks & relics);
- Timetables and other information on shows, deployments, exchanges and other aviation events;
- Radio Activity (new frequencies, call signs);
- Show reports (full reports in all major aviation events);
- Fokker news (all about Fokker aircraft, including the Fairchild F-27 and FH-227);
- Full coverage of the Dutch Civil Aircraft Register;
- Trip reports from all over the world;
- A mix of large and small, civil and military articles.

If you would like a subscription, or more info on our magazine, please check out www.scramble.nl/subscribe.htm or send an E-mail to subscribe@scramble.nl

You are welcome to visit the official website. The website is in English and free for everybody. You can find more information about **Scramble** in the "Magazine" section of the Internet site. As **Scramble** Magazine covers both civil and military aviation, we have created sections for every interest. For a growing number of countries you will find an extensive Order of Battle on the site with unit-badges, database, base-overview, maps, pictures and links. For a considerable and growing number of countries you can access our database for your own reference. Scramble-subscribers even have more privileges and can get more information out of our databases. We hope you will enjoy our site. The pages are updated on a regular basis, so come back often to our website!